The Recycle-to-Land Series

REVERSING GLOBAL WARMING FOR PROFIT

If it isn't financially sustainable,
it isn't environmentally sustainable

* The inescapable route to limiting the damage
* *Just common sense*
* *Pollution control bonus*
* *Farming can deliver food and biofuels*
and do it now

Bill Butterworth

"It has happened before, so we can
do it again!"

1

Paperback ISBN 9781904312819

Published in the UK by MX Publishing
335 Princess Park Manor, Royal Drive, London, N11 3GX
www.mx-publishing.co.uk

Cover Design by www.staunch.com

This book is dedicated with great respect
to gardeners, farmers and foresters,
the men and women
who made the countryside what it is,
have maintained it for a thousand years,
and
who will hand on the heritage of the land
in a fit state
to our children and our children's children.

"An academic is concerned with pushing knowledge to the limit of knowing.
A technologist is concerned with pushing knowledge to the limit of doing."

This book does not set out to be a substitute for a PhD degree. It is about a combination of science and technology, and of welding good farming practice with recycling wastes and pollution control, and with business development, in order to build what a reasonably sensible, science-based observer can accept as a logical and practicable way of solving some problems and making progress towards widely desired gaols.

It is acceptable simply because it is also a historical record; it has been done; done safely and sustainably from an environmental point of view, and it is financially sustainably, too.

This book is, then, logically about why three things are key to the survival of our children and maybe ourselves;
-         arresting population growth,
-         arresting the burning of fossilised fuels and replacing them with biofuels,
-         the technology and practical ability of taking Carbon dioxide out of the atmosphere and pumping Oxygen back in on a global scale, while producing enough food to feed everybody.

We must   *Stop - Think - Do It Now*

CONTENTS - *(pages in brackets)*

5

Chapter 1
Introduction
Not About Global Warming
Sustainability

Not About Global Warming
If you don't believe that global warming is an immediate threat, not just to our children, but to ourselves (whatever your age), then you are to be admired! You are free from the concerns of any thinking and informed, reasonably well educated person. You are well advised *not* to go and read Sir David King's excellent book; "The Hot Topic"[1], nor watch Al Gore's video "An Uncomfortable Truth"[2], nor read the United Nations' commentaries[55] on global warming on climate change and its speed of progress. Live in ignorance and be happy.

If you think that global warming is occurring but it is just a natural phenomenon and nothing to do with human activity, you might have a point. You might read Gavin Menzies outstanding, well-researched book[3] on "1421 – the year China discovered the world". He points out that since 1421, sea levels have risen by "between 4 and 6 feet" (1.2 to 1.9m). Translating that observation into this text, that could not have been due to human activity all the way back in 1421 because then we were not burning fossilised fuels. Sorry, the rise in sea levels did not occur back then, it has occurred in the last 100 years. If the same thing happens in the next 100 years, then a historian looking back will be able to trace what Menzies observed; a further, significant, mapable change in what is left above sea level. While there may well be a natural change, nothing to do with burning fossilised fuels, the fact is that there is no doubt whatsoever that most, if not all, is down to human activity. In any case, it is happening and we need strategies and the practical ability to at least slow it down.

There is no doubt about global warming. There is no doubt that burning fossilised fuel reserves is a key issue.

This book is not about global warming but in the consideration of how global warming can be slowed down, stopped at this point and, preferably, reversed. Any solution has to be seen in the context of a changing world environment with a complexity of political and economic changes affecting the environment which is, itself, dynamic.

"The environment" is global (big), intricate (complicated) and what scientists call "dynamic" (living and continually changing). It is always changing, with every organism, every situation, reacting with its neighbours and the wider conditions of its situation. It is impossible for any individual, any situation to be static. Everything that happens at a point affects everything else. On the face of it, that makes the search for control over the global climatic change apparently utterly beyond our influence, never mind control.

Despite that apparent despairing view of the future of the human race, there are some basic issues which can be tackled and can be delivered provided we stop discussing it and get on with three basic things.

This book is about three necessary prerequisites for the slow down and arrest of global warming. We may even be able to reverse it a little if we go down the route described here. Two out of the three prerequisites are raised and put in place without detail. There is no apology for that as others will detail those elsewhere. What can be done here is show how the third can deliver what is necessary, *provided* the first two are in place and running alongside.

Everyone agrees that there is no "Magic Bullet", a single right solution, one route to salvation, one size fits all, and, logically, that must be right. There can be no one energy source that fulfils all human needs and does not destroy our current environment. However, there is one guiding principle which is fundamental and embraced usefully in one word: *sustainable.*

Sustainability
The One Best Way
Benjamin Franklin, one time President of the USA, used to use the expression: "The one best way".

Never mind about whether something is "renewable". Never mind if it can be labelled as the latest fashionable "environmental" fad. That "one best way" which we are looking for is to ask a question and be honest about the answer. If I do what I am doing now, and carry on doing it, will I still be able to do it just the same in a thousand years? Will the results, then, of what I am doing be just as good or better? Will I be able to do it for another thousand? *That* is what sustainability is about.

There is an old Apache saying: "The land is a mother that never dies." Well, there is a remarkable truth about land; they've stopped making it.

Chapter 2
Uncomfortable Choice Number 1
Population Growth and Control
Population and Global Warming
Population and Energy Use
Postponing Death

Population Growth and Control
Before looking at population, there is something that is worth remembering about what people stand on – Land; and they really have stopped making it. Indeed, if sea levels rise as predicted, we are going to have significantly less of it.

If you look at almost any report on current world population figures, it will look something like Figure 2.1 below.

Figure 2.1

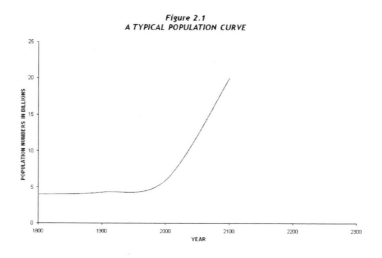

Figure 2.1
A TYPICAL POPULATION CURVE

We can be fairly confident of the date and, for the numbers, there are some reasonably reliable figures on world population. For the logic of this argument, it does not really matter whether the population is taken as six billion or even seven. By the time you read this, it may be 7.5 or 8, or beyond.

Figure 2.2 shows a typical, idealised, stable population curve which levels out at a "comfortable" maximum and stays there.

Figure 2.2

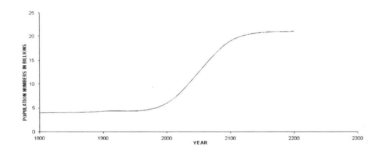

Figure 2.2
A TYPICAL, IDEALISED POPULATION CURVE

Everyone assumes that is what it will be like and then argues about at what point it will level out, or avoids the unpleasant conclusion that it will level off and assumes that we can keep on going. Only an idiot would argue that there is no limit; ultimately, there has to be a point where, however efficient we are at meagre living, at efficiency of resource use, the population will outstrip the available resources and level off. Most reasonable discussion appears to argue that it will level off at 15 to 20 billion sometime in the next 50 or 100 years. The truth is that such estimates have a number of weaknesses. Nevertheless, it is a nice picture and it makes us feel comfortable, partly because we feel we don't have to face the process of how it levels off, or whether there will be some other happening at that point. Do populations "level" off smoothly to a stable situation and are any of the figures of 15 or 20 billion a reasonable guide? How do we know that it will be 100 years off? Might it be 50 years? Might it be tomorrow morning? How will it level off? What will be the process?

A key to the answers to these questions can be seen in William Stanton's thorough study of world population curves[4]. Nearly every country or regional population in the world has hit this exponential rise. Generally, birth rates have stayed high and death rates fallen. It is certainly true that many developed economies eventually show a stabilising population because birth rates fall. Indeed, many argue that this will solve the problem in that every region of the world will follow the same pattern and that world population will stabilise and the application of medicine and economic wealth proceed to levels which are, at least in part, predictable levels. There is plenty of evidence for this in the academic and government literature on population.

The idea that *every* region and country will ever do this does, logically, seem very unlikely. We do have the technology, in place right now, to actually feed everyone in the whole world but we don't for a complexity of reasons which are political, to do with military power, and logistical difficulties in equalising distribution of resources. The latter reason will always apply simply because of the protective instincts of the "have" nations, versus the "don't have" nations.

Similarly, there will not be an evening up, or rather down, to stability, of population growth in all nations; it is not the nature of things. However, suppose for the sake of argument, that there comes that point of stability in world population as in Figure 2.2. What will it look like and will it stay like that? How will it affect greenhouse gas production?

Military Wars
One of the characteristics of starving people is that they do not have the resources to go to war to gain food. Not, that is, military war in the modern, high-tech military sense. They do not have the resources to develop and manufacture armaments. Well, even that is not quite true in that sometimes very poor nations have dictators who develop military power. However, the point is valid; generally speaking, the rich nations have the wealth to prevent military attack by poor nations.

Where history shows that wealthy nations have a weakness is in guerrilla warfare or, weaker still, in "illegal immigration". Large countries have large borders. Large borders are difficult to police. It is therefore inescapable to conclude that there will always be pressures to even up apparent wealth.

All people, even those actually dying, use energy. Wealthy people use a lot of energy, even if they live in a warm climate and don't use a motor car or fly in an aeroplane. High density populations use more energy per person because of the energy cost of logistics to move food and goods, sewage and water, and also because of the increasing demands of health needs and disease control. All our plastics, metals and just about everything we use for the sustenance of life is energy dependent and most of that energy comes from burning fossilised fuels. Even the populations of the developed countries of the world surfing the net is dramatically increasing energy consumption.

There is no escaping it, survival demands energy. Currently, the human race is geared up to using fossilised fuels to find that energy.

Does population collapse actually occur?
One of the most interesting curves Stanton[4] draws attention to is that for the British Isles over the last 1000 years or so.

Figure 2.3

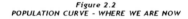

Figure 2.2
POPULATION CURVE - WHERE WE ARE NOW

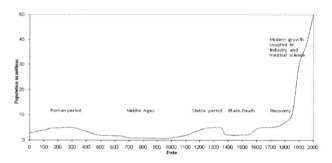

There have been catastrophic collapses of population more than once in British history; three from the Black Death alone. This one disease, in a series of pandemics, spread over a comparatively short number of years, more than halved the population of Europe. There are some who take the view that HIV will/is, doing the same in Africa. Will HIV be controlled? Of course - well probably - but the cost? When? How many will die

14

before it is under control? We don't know. The point is that catastrophic falls in population do occur. They occur because population pressure pushes the available resources to the limit of survival and then there is an adjustment. This *always* happens; it is the nature of population. Human populations are different, of course, as we exercise a much greater control over environment than any other species. (Well, we think we do!) We are capable of pushing frontiers back. All of this is true and we will keep pushing the frontiers back. Is there a limit? No, but as population grows, then catastrophic collapse becomes an increasingly likely risk. There are many of these risks but there are two which are worth further thought in order to demonstrate that this is a real problem. Firstly, population density has a remarkable consistent effect on the social habits of every species. Put chickens too close together and they peck the weaker ones to death. Put too many pigs together and they become cannibals. There is no reason to believe that the human species is different. Indeed, there is plenty of evidence to confirm that we are exactly the same. People in cities become more aggressive and self-protective, crime rates go up in unstable communities; it is inevitably the nature of survival.

Secondly, as population increases in numbers and the available space per individual declines (as in growing cities), then it becomes more vulnerable to disease. As population densities rise, resources become stretched, pollution increases, then the likelihood of mutations in diseases increases, the speed of infection passing between individuals increases and the progress to catastrophic collapse increases.

Logically, the likelihood of catastrophic collapse of global human population is unavoidable. There is, of

15

course, argument about what the truly sustainable level of population will be. Suppose, for example, it is in the region of 3.5 billion. Then collapse will overshoot that figure and oscillate until it stabilises and then, over the next thousand years or so, the whole process of population growth will start again – we hope.

Figure 2.4

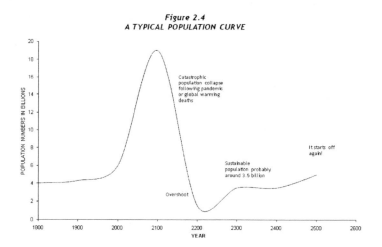

Figure 2.4
A TYPICAL POPULATION CURVE

Population and Global Warming

So, where is this population discussion taking us and what has it to do with reversing global warming?

Firstly, population numbers are a real issue. In a recent debate in the UK parliament, David Kidney MP raised this issue with respect to the UK population. He was quoted by Martin Livermore[52] in a piece for the Adam Smith Institute and he also quoted the Optimum Population Trust as saying "UK should be limited to 30 million" (at the time, the UK population was approaching 70 million), and The Global Footprint Network tells us that

16

we (the world) are currently using the resources provided by 1.3 planets. However, Kidney went on to say "It's important, in my view, to keep a sense of proportion and to look at population size in its context alongside other considerations, including developing the means to provide a larger world population with the water, food and energy it will need."

It may be that population will peak and then collapse to maybe only a couple of billion before it stabilises at whatever figure, maybe 3 or 4 billion, maybe 1 billion[4] - or whatever figure, but it won't be 20 billion or whatever the peak turns out to be. Populations do not stabilise at their peak; they *always* collapse back to a lower level. The logic confirmed by population studies indicates quite conclusively that stable population levels are a fraction of the peak. When the collapse happens, and it appears logical that it will, then it will be painful. It will be painful if you are one of the ones that die. It will be painful if you are one of the ones who survive because the world will be a very different place after that fall. Many of the things we take for granted will not be there anymore; the fact that the human race used to have the knowledge of how to make a garden spade does not mean you can have one. Could you go and make steel on your own?

Now, if you are a politician, perhaps not a good idea to be in charge when the collapse becomes obvious as imminent. Better to be prepared to try and be one of the nations that planned to avoid it.

Secondly, there is something perverse about any catastrophic fall in global population; we will dramatically cut greenhouse gas production.

So, it may be that catastrophic collapse of population due to disease will cause a collapse of numbers in the human race before global warming makes the world too difficult for current population levels. Either way, there will be a fall.

Until that point, what has this to do with global warming? The answer is about energy use.

Population and Energy Use
It is worth going back to Al Gore's summary of where we are. Gore[2] did not get his Nobel prize as a trivial passing remark by a group of uninformed idiots. This is serious stuff beyond reasonable objection; there is a serious problem here. There is no doubt that the rise of human populations is occurring. There is no doubt about the increase in the use of energy from all sources. There is no doubt about the increase in greenhouse gases. While there are many different greenhouse effects and many gases worse in their warming effect than Carbon dioxide, the fact is that Carbon dioxide emissions from burning fossilised fuels is a major issue; some would argue *the* major issue.

Conclusion 1
Whether we like to face it or not, population growth is an issue. We ignore it at our peril. It is nothing whatsoever to do with race, it is to do with numbers.

However efficient and sensible we get in the use of energy, the fact is that humans use energy. Every extra human individual makes a contribution to greenhouse gas production and global warming. It may be possible to have a large population and avoid that, but it appears logically to be unlikely, certainly in the timeframe that global warming is likely to be catastrophic.

So, population is an issue. In the short run, every new human means we burn a little more fossilised fuel and produce a little bit more Carbon dioxide.

Conclusion 2
Postponing Death
Logically, all the compassion, all the charitable care, all the medical science, they never save lives, not ever. What they do is postpone death and, hopefully, make it less uncomfortable. Therefore, the exponential rise of population seems more likely than not and catastrophic collapse (probably due to a pandemic of a mutated virus disease) an inevitable consequence.

There is a outstanding question; If we lack the global political will to stop unfettered population growth, as seems likely, will there actually be a catastrophic collapse and will this, then, solve the issue of humans' contribution to global warming? Would such a collapse stop a catastrophic rise in global temperatures? Probably not; it would be too late.

Chapter 3
Uncomfortable Choice Number 2
Reduction of Fossil Fuel Consumption

Reduction of Fossil Fuel Consumption

Ruminants are animals such as cows and antelopes, sheep and goats, buffalo and Gnu, which have a rumen – a large stomach full of micro-organisms which allow these grazing animals to digest cellulose (grass and other green material). The process in the rumen is "anaerobic" which means that the micro-organisms do their work with little or no Oxygen and, in the case of these animals, they produce Methane. Methane is a "greenhouse gas" with a far more potent effect than Carbon dioxide. And there are billions of ruminants on the world's surface. There is some really exciting work being done by Australian agricultural scientists[5] to see if the micro-organisms in the stomachs of kangaroos (which eat and digest grass but do not produce Methane) could be transferred in some way to ruminants. It has a good chance of success but, even if it does, it may take hundreds of years to change the world's population of ruminants.

There are many other contributors to climate change. It is not the purpose of this book to list and evaluate the relative contributions of each. What is already clear is that we have to tackle all of these as quickly as possible but it remains inescapable that burning fossilised fuels is a major contributor, most would argue, *the* major contributor.

Consider Figure 3.1; plants take Carbon dioxide out of the atmosphere and water through their roots and produce small sugar molecules. The most commonly quoted is the six-Carbon sugar. When animals (including humans) use muscle energy, they push the equation the other way and we get the Carbon dioxide and water back. Keep those two equations in balance and we have long term sustainability.

The next line in the row of equations shows the burning of a small Carbon molecule, propane, and how much of each element is used up.

The final equation is the burning of a large Carbon chain molecule, actually found in petrol. This is what happens when a petrol-driven car is run.

Figure 3.1

Figure 3.1
The Basic Equations in Managing Real Sustainability

Energy from the sun

Plants take CO_2 and water ⟶ To make large Carbon molecules

$$CO_2 + H_2O \longrightarrow C_6H_{12}O_6$$

⟵ Animals and incineration push this the other way

The balanced chemical equation reads;

$$6\,CO_2 + 6\,H_2O \longrightarrow C_6H_{12}O_6 + 6O_2$$

Burning a small Carbon molecule reverses this process;

$$C_3H_8 + 5\,O_2 \longrightarrow 3CO_2 + H_2O$$ - plus some energy as heat which we could use for making electricity
Propane   Oxygen

Burning a big Carbon molecule would read;

One molecule from Petrol $$2C_{36}H_{74} + 109\,O_2 \longrightarrow 72\,CO_2 + 74\,H_2O$$

Rounded figures   1 kg + 3.5 kg ⟶ 3.2 kg + 1.3 kg

The green leaf can push this equation in the opposite direction.

Never mind the Carbon dioxide, where is the Oxygen going? The only reversal mechanism we have right now is the green leaf.

It is pretty clear that the burning of fossilised fuel reserves has to slow down and, preferably, actually stop

21

as soon as possible, if not now. Tomorrow is too late. That, however, is not going to happen. Even if everyone agrees the principle (which they won't), they will then have to agree a timetable (which they won't), and then every nation will argue if this is a special case (which they will), and action will be too little too late (which it will be). Nevertheless, the fundamental need is to slow-down or stop burning fossilised fuel and this must happen and soon. We have to make a start.

Of course there are alternative sources of energy. Some are "free" and sustainable long-term. Many are not. It is important to distinguish between "renewable" and "sustainable".

Hydro-electric power is potentially sustainable. Unfortunately, wind power is often not. (Wind turbines take much energy to make them, in many situations are idle too much of the time and, if the wind fails, then existing/other energy-source power stations have to be used – so they cannot be decommissioned.)

Wave and tidal power are potentially sustainable. Unfortunately, many wood-powered Heat–and-Power plants are certainly "renewable fuelled" but are often not sustainable. (The energy needed to harvest/collect the wood or biomass, remove any water if present, chip or shred, transport to the facility, and remove the ash, may not be attractive from an energy balance point of view).

And so on; the point being that whatever "renewable" energy is being considered, the real, in-practice, energy equations need to be examined within an honest framework. The ultimate question is; if we do this for 1000 years, will we be better off or worse off?

Motor Cars and Electric Engines

We really do need to be careful about electric cars and "hybrids" (using a petrol or diesel engine coupled to an electric motor and batteries). A car run on an electric motor may be, in itself, "clean" and good from the point of view of local pollution. However, where the electricity comes from, often from a remote power station many miles away, generating its fuel from burning something, probably fossilised fuel and probably with a very poor total system efficiency, is hardly a sustainable solution. It often is far from it. The energy cost of the batteries and what to do to dispose of them, or recycle them, at the end of their useful life, is another consideration which is often over-looked and likely to be a negative.

Internal Combustion Engines

There is a real problem at the end of the discussion about alternatives; it is the population of internal combustion engines. The starting point of all action is: we are where we are. It is sometimes said that there is a rat for every human on the face of the earth. Well, there is probably about one internal combustion engine for every human on the earth, too (more in developed countries). Whatever the relationship to global human population, global engine population is certainly a very large number. Whatever that number is, the industrial capacity available or conceivable must imply that it would take many years, several human generations, to build and commission a replacement of all those engines. Indeed, it is logical to assume that any solution might involve so much resource use so as to be counter-productive in, at least, the short term. The implication, logically, is that we have to use what we have got in the short run and that means use the existing engines. But logically, if we are to arrest global warming, we need to find a sustainable, liquid fuel to power them. That necessarily means biofuels. However, not all biofuels are sustainable.

Chapter 4
Uncomfortable or Comfortable Choice Number 3
Soils and Greenhouse Gases
Biofuels from Crops Grown from Wastes
- The Ultimate Mimic of the Carboniferous Era
Reforestation

Soils and Greenhouse Gases

The question is often asked about individual processes; is it really "green"? The answer is that there is nothing in the real world that is all good or all bad. There are no clear distinctions between "black" and "white" answers. Some processes are, on balance, sustainable and some, on balance, not so. All biofuels have at least one advantage and that is that they are replaceable and could be generated anew. However, many are only replaceable at a cost which is not sustainable.

For example, burning poultry litter in a factory to produce steam to drive a turbine to produce electricity does actually work and the electricity is a "renewable fuel". However, if the poultry litter previously was put to land as a fertiliser, then, presumably, the fertiliser value will have to be replaced. Nitrogen fertiliser is made by passing electricity through a large electric arc (with the electricity likely to have come from burning fossilised fuel) and it is likely, very likely, that the energy cost of replacing that fertiliser value will be several times what the factory produces as electricity. Further, the Nitrogen in the straw must, presumably, go up the flue of the furnace and come down somewhere else as acid rain. Not sustainable. Not common sense but it has happened.

Another example is the conversion of part of the input of coal-fired power stations to burn straw. One factory in the south of the UK is reputedly so converted to take 30,000 tonnes of straw per annum. The energy cost just of one set of contractor's equipment (and there will be several) of baling up the straw in the field and transporting to the power station, will involve something in the region of 400 to 500 horsepower (hp) of engines driving the machines involved, all using diesel, for 3 or 4 months. The engines to shred the straw and feed it into the plant will involve another 500 to 1000 hp for, presumably, 12 months of the year. Just possibly, maybe probably, it would be better to put the diesel fuel used to drive all of that direct into the power station, get around 1500 truck journeys off the road and leave the straw in the field. If it is ploughed into the soil, then it produces a Carbon sink which can be built up. In fact, the Carbon locked up in that sink does oxidise to Carbon dioxide and go back to the atmosphere. However, it happens relatively slowly and is partly dependent on the method of cultivation so that the oxidation may be at as much as 35% in high power input and conventional ploughing systems of cultivation, down to 10% per annum in direct drilling (sometimes called zero till) cultivation systems. That degradation is built up with a new 100% of each years' harvest. One of the problems with asking power stations to burn "renewable fuels" is that the burning still turns fuel into Carbon dioxide. When the straw (which the farmer would have ploughed in and locked up the Carbon) is burned in the power station, the Carbon in the straw is immediately turned into Carbon dioxide.

The key question is: on balance, is the proposed operation sustainable? "Sustainable" means that we can carry on doing this activity, not just for a thousand years but ten thousand and beyond. The only human activity which has been going on for that length of time and on a big scale is farming. So, what has farming to offer to the control of global warming?

In farming, there are two key issues. Firstly, that the green plant can take Carbon dioxide out of the atmosphere and pump Oxygen back in. Secondly, the Carbon-based organic matter in soils does oxidise (at varying rates) and "leak" the Carbon back as Carbon dioxide or, worse still, Methane, and the Nitrogen back as Nitrous oxide (another, very potent greenhouse gas with over 300 times to warming effect as Carbon dioxide). However, these leakages back into the system are relatively small and can be controlled to some extent. As Sara Wright's researches[6] showed, and others before and since[7], the rate of leakage is dictated by a number of factors including soil temperatures but maybe the major one is under the farmer's control and that is the violence of the cultivation. High power input generally increases the rate of oxidation of organic matter, producing both Carbon dioxide and nitrous oxide. The greatest effect results from the use of pto-driven implements to "force" a tilth. ("pto" means power-take-off and applies to implements which use the power of the tractor engine in a direct drive to rotate or oscillate implements to "force" a tilth.) Research results show a wide range of emissions, from as much as 1.5 % to less than 0.5%, of Nitrogen back to atmosphere as nitrous oxide. Conventional cultivation systems, involving implements such as the mouldboard plough, tined cultivators and power harrows, in several passes, will cause relatively rapid oxidation of organic matter and maybe 35% per annum of the Carbon in the organic matter will be lost on a declining basis [8] [9] [10], i.e. 35% of what's left will be lost next year, and so on. However, with direct drilling, or zero till, that figure may drop to only a 10% loss.

Biofuels from Crops Grown from Wastes - The Ultimate Mimic of the Carboniferous Era

Consider the following figures then follow the argument through. Take special note of the size of the Carbon dioxide box, top centre, of each Figure. Bear one other thing in mind: the individual figures are put here just as an example. It is quite possible to put different figures because of different crops, different yields, different views of how much Carbon a particular material holds and so on. What cannot be dismissed is the principle: this is a sustainable loop, it is the only one shown, to date, to deliver that sustainability which we seek. The evidence for that sweeping statement is simple and inescapable; it has been done before, 360 million years ago.

Figure 4.1

Crops harvest sunlight by taking Carbon dioxide out of the air and water via the roots and releasing Oxygen. If the crop is burned in air, we use up the same amount of Oxygen and release the same amount of Carbon dioxide. On the face of it - a balanced and sustainable equation.

Figure 4.1
CROPS TO BIOFUELS - THE BASIC ROUTE

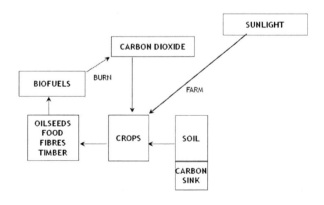

Figure 4.2
One of the traps which biofuel production easily falls into is not to make sure that the feedstock it uses has an attractive Carbon footprint. If the feedstock crop is grown with mineral fertiliser, it must be remembered that that fertiliser, where it is Nitrogen fertiliser, will have been made by passing air through a large electric arc. The electricity is likely to have been made by burning fossilised fuels. Nevertheless, Yara, the Norwegian manufacturer, has shown very significant improvements can be made in process efficiency and production of emissions.

Figure 4.2
CROPS TO BIOFUELS - MINERAL FERTILISER

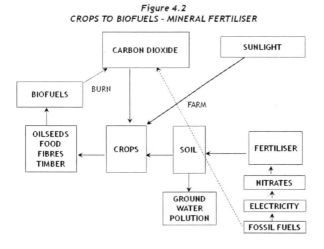

Figure 4.3
Biofuels, from crops, from compost, made from wastes is a different situation. It is not only Carbon attractive; it is Oxygen attractive, too. Also, when a crop produces oil seed which is used to produce the biofuels, the rest of the crop (the leaves, stems and roots) will help form a Carbon sink with the Carbon left in the compost.

Figure 4.3
CROPS TO BIOFUELS - FROM "WASTE"

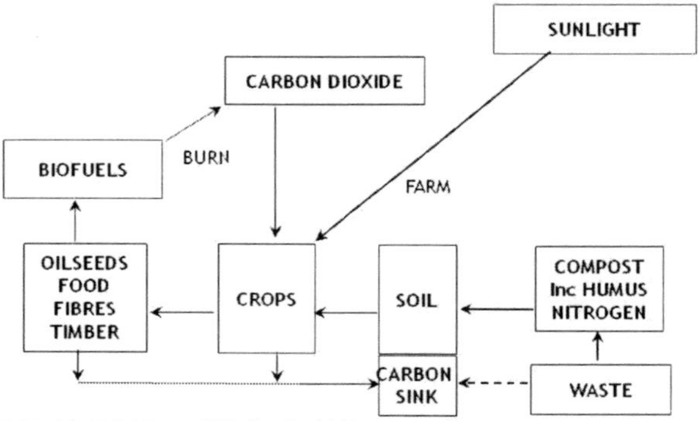

1 ha can produce 1 tonne of biodiesel which, when used, gives 5 tonnes of CO_2. PCCSS, then, captures & stores in the soil a net 74-5 = 69 tonnes of CO_2 per ha. It also releases about 73 tonnes of Oxygen.

Building the Carbon Sink
Much has already been made in this text of the opportunity to fix Carbon and reverse global warming. It is also worth hammering on about biodiesel and PPO - Pure Plant Oil. Biofuel production from crops has a major global attraction *provided* it is done using wastes to fertilise the crops. It also, maybe just as important, can be done on a large or small scale and done locally. It is possible to limit trucking to collect wastes on a proximity basis, grow and harvest oil-bearing crops in that locality, convert to biodiesel on the same local operation and only export the surplus diesel production. If the whole operation is carried out on a community

30

basis, the energy equations do stack up and there is a real gain in Carbon capture. It is important to observe, however, that the key to success is not CCS (Carbon Capture and Storage – which is being talked about with respect to limiting the damage caused by burning oil and coal) but in PCCSS (Photosynthetic Carbon Capture and Storage in Soils) using photosynthesis and soil. If PCCSS can be used while significantly cutting the burning of fossilised fuels, then there is a very distinct possibility of reducing and reversing global warming[12].

Plants harvest sunlight. A plant uses the energy from the sun, with the help of the chlorophyll in the green leaf, to take water (Hydrogen and Oxygen) via its roots and Carbon dioxide (Carbon and Oxygen) via its leaves from the air to make sugars and then oils. There is a bonus because there is some Oxygen left over and that goes back into the atmosphere. In a balanced ecological system, plants and animals circulate Oxygen and Carbon dioxide in this way. If we chose oil-producing crops, then we can use that oil to produce biodiesel. If the process can be done locally, then the logistics and energy consumption involved in trucking to centralised processing and distribution can be dramatically cut. Figures from decentralised operations in the UK[8] indicate that this reduction in tonne-truck-kilometers (used as an indicator of the energy cost of taking one tonne on a truck for a kilometre (if it is preferred to be imperial, tonne truck miles can be used) can be in the region of 65 to 85%. If the crops can be grown with fertilisers made from wastes (usually by composting and/or mulching), then it is possible to eliminate the energy involved in industrial manufacture of mineral Nitrogen fertilisers which use significant amounts of electrical power (usually generated by burning fossilised fuels). There are many local wastes which can be used to produce a choice from over a hundred oil-producing crops.

The Carbon Dioxide Figures
The figures given here in this text are from a UK research and development project but similar figures will apply elsewhere and with other crops. These figures are related to oil seed rape and the wastes are separated municipal biodegradable wastes, MDF from furniture production/recycling and some industrial wastes. See Figure 4.4 with the explanation which follows. The figures used are an actual field example which might be duplicatable elsewhere and are a reasonable guide. It is quite possible to argue with the detail of the figures but not the principle.

Figure 4.4

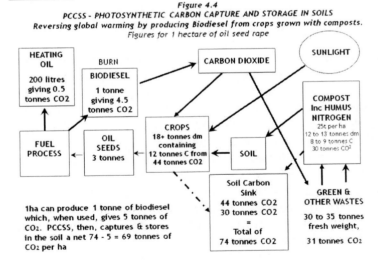

Figure 4.4
PCCSS - PHOTOSYNTHETIC CARBON CAPTURE AND STORAGE IN SOILS
Reversing global warming by producing Biodiesel from crops grown with composts.
Figures for 1 hectare of oil seed rape

The Figures on Carbon Capture and PCCSS
One hectare of oil seed rape crop will produce around 3 tonnes of oil seed rape and that will yield about 1 tonne

of biodiesel. In producing that 3 tonnes of seed, the crop will also have produced probably more than 5 to 10 tonnes (say typically 7.5 as an example) of leaf and stem. Generally, most crops will produce as much dry matter below the ground as they do above the ground. Therefore, the total dry matter production might be in the region of 18 tonnes per hectare. Now, that 18 tonnes of dry matter is mainly Carbon and will contain about 12 tonnes of Carbon. That Carbon will have been taken out of the atmosphere by the green leaf chlorophyll process. As the atomic weight of Carbon is 12 and that of Oxygen is 16, and there are two Oxygen atoms to each Carbon in Carbon dioxide, the original 12 tonnes of Carbon in the dry matter will have removed 44 (12+16+16)tonnes of Carbon dioxide out of the atmosphere. However, that is not the end of the capture. (Remember that 44 and carry it forward.)

If the fertilisers are made from biodegradable wastes, then those wastes also contain Carbon which will have come from the same green leaf process. Compost made from green wastes would, in the Land Network R&D programme, be likely to loose around a quarter of its weight during the process. Regulation in the UK will allow compost to be applied to the crop at around 25 fresh weight tonnes per ha. To produce that 25 tonnes would take maybe 30 to 35 tonnes of green waste which would have taken around 31 tonnes of Carbon dioxide from the atmosphere.

The 25 tonnes of compost used on each hectare would contain probably 12 tonnes of dry matter and, therefore, around 8 tonnes of Carbon, which would have been fixed by removing a bit more than 30 tonnes of Carbon dioxide out of the atmosphere. (In fact, the original green waste would have had a little more Carbon in it, say 31 tonnes, but a little is lost back to Carbon dioxide released to the atmosphere by the composting process. So, the total removal of Carbon dioxide from the atmosphere is the 30 tonnes in the compost which is locked up into the soil "Carbon sink", plus the 44 tonnes in the crop. That makes this programme remove a net 74 tonnes of Carbon dioxide from the atmosphere to produce 1 tonne of biodiesel. When the biodiesel is burned, and its co-product bioglycerol which will also be produced by the biodiesel production process, there will be 5 tonnes of Carbon dioxide released back to the atmosphere. The net, then, is 74 minus the 5 equals 69.

Global Warming Reversal

The conclusion, taking it all into account, is that this process, on these figures, will remove a net of 69 tonnes of Carbon dioxide from the atmosphere and release around 73 tonnes of Oxygen back for us to breathe. The reverse would have happened if this material had been burned in an EfW (Energy from Waste) plant, which incinerates waste to produce heat, which can be used to produce steam to drive a generator, to produce electricity. However, if the Figure 4.4 process were used with the same waste, through the green leaf, burning the oilseed rape oil in a diesel engine, direct coupled to a generator would produce more electricity *plus* the 69 tonnes of Carbon dioxide removed from the atmosphere and 73 tonnes of Oxygen put back in. There is no way EfW can get anywhere near these figures. What EfW would have done would be to turn the Carbon in the wastes back to Carbon dioxide - back to the 31 tonnes right back at the beginning of this discussion. Comparing these as alternatives, the compost route is 69 plus 31 equals 100 tonnes better at removing Carbon dioxide, gives us back the Oxygen and delivers more energy. This route really is sustainable. The detail of the figures can be argued about but they are not fundamentally misleading. The principle is sound.

Clever? Not really. It is just a question of mimicking what we observe in nature and how it managed to create ecological balance in Carbon management. In the UK, there has, at peak, been around 450,000 hectares of Set Aside land under EU rules on limiting arable production. When that was set up, nothing could be grown on that land. After some years, the EU, in its great wisdom, allowed to grow energy crops on that land. Then, Set Aside was dropped. At 44 tonnes of PCCSS that land, could have locked up, every year, over 20 million tonnes of Carbon dioxide (24% of the Kyoto Protocol estimate of total UK emissions) and produced 450 million litres of biodiesel with less than 1 million tonnes of Carbon dioxide produced when actually used.

These figures suggest we could reverse global warming. Well, it is a bit more complicated than these figures suggest, of course.

Firstly, there is energy used up in process logistics. This is certainly helpfully lower if the process is based on community operation. Secondly, there is leakage of Carbon dioxide from PCCSS systems at maybe 20% in conventional cultivation systems and maybe as little as 10% in reduced cultivation or "zero tillage" (called "direct drilling" in the UK) systems. Reduced cultivation marginally increases the emission of another greenhouse gas, nitrous oxide. However, there is a balancing factor which is that the method described here saves very large amounts of energy which is used and transporting mineral fertilisers which consumes very large amounts of electricity. So, these figures are a simplification for the sake of presentation here. However, they are not misleading. This could be *the* delivery mechanism, maybe it is the *only* delivery mechanism, which will work. This is a global scale system which already exists.

Technology can deliver management solutions but only if there is a political will to provide the wider framework to accommodate a raft of measures. Firstly we have to dramatically cut fossilised fuel burning and quickly. We need to nurture those reserves. One idea here is to fund the United Nations by using it to administer a global aviation fuel tax and we need the USA and China to seriously sign up to Kyoto principles. Fortunately, under President Obama, the USA tide has turned. Secondly, we have to organise global waste management to re-direct it into safe land management for energy. Thirdly, we have to study and manage the effects of the potential loss of global dimming. Lastly, and by no means least, we have to consider that much of the cash which could drive the political framework to make these things come right can only come from getting the oil companies to buy into the framework. They could start by thinking about looking at the centralised production of bio-ethanol and about distribution of that bioethanol and the surplus of biodiesel from community production from wastes. Delivery is possible but the clock is ticking faster than ever. Remember that the argument here depends on producing those biofuels from crops that are fertilised not from mineral fertilisers manufactured specifically for the purpose, but from wastes.

Algae Production and Biofuels from Algae
The oil companies in particular have become increasingly interested in algae, such as Cyanobacteria and Spirulina, simply because they are green plants and they may be manageable in liquid systems which can be mechanised explain in one short sentence what algae

36

could be useful for in oil industry. In theory and in trials, it is possible to produce fuels from algae. They will also do what the green leaf in farming does; remove Carbon dioxide from the air and pump Oxygen back in. It is not the purpose of this book to examine the possible route for culture of algae to lead to liquid biofuels. The researchers, however, may draw from agricultural experience. We know from high-tech inputs into the genetic manipulation of crops such as maize, coupled to carefully designed cultivation, that we can get close to total utilisation of the available energy from the sun by restructuring the way leaves grow so that they can "mop up" all the available light. If we have a layer of algae, the same rules apply; there is a limit to how much sunlight falls on an area and algae don't grow and fix the sun's energy in the dark. There may be a very long way to go before these systems can be managed.

Algae are certainly interesting and worth pursuing. For the here and now, however, we know that we can reclaim deserts and uplands with "wastes" and produce both biofuels and food by relatively well-established normal agricultural techniques. We know they are safe. We know they work.

Other Alternative Waste Treatments and Biofuels
The example above, of waste into compost, fertilise an oilseed crop and produce biodiesel on the same farm has happened. The figures can be argued about, but not the principle. The basic principle is simple enough; use wastes to fertilise a crop which involves chlorophyll (in a green leaf or algae) to convert sunlight to a usable energy source.

The waste treatment can be direct to land, it can be composting, it can be anaerobic digestion, pyrolysis or anything else which we already know about or may come up brand new. However, the bit that matters is sunlight on chlorophyll. We know this works as a mechanism. We know how to manage it. It can be used globally. It has been done before.

Various estimates have been made of how much land would be necessary to produce enough biofuels for the world's cars, trucks or aeroplanes. Two thirds of the globe is covered by sea and we are a long way off managing that with green algae production and the potential management risks of using this route are very significant in terms of pollution risk simply because of tides, winds, currents and water turbulence.

The safe route is land. The bad news is that "they have stopped making it". The good news is that there are large tracts, very large tracts, which are not cropped. However, much of these areas are difficult to farm for many reasons and much is desert.

Reforestation
Could we really grow trees on the tops of the UK's highest mountains and make the Sahara desert bloom? Yes we can. We know this because 5000 years ago the whole of the UK used to be covered in trees[12] but they were cut down to build ships, the land was grazed for wool production and over-grazed; the result was the washing of soils into the sea. That land could be reclaimed with soil based on wastes and re-forested. That could be done on land up to 2500ft (770m) of UK mountains and on the sands of the Sahara.

Is there any evidence that reforestation would affect, possibly have a reverse effect on, global warming? Well, there is some logic that it might because trees can absorb up to 1.5 times the Carbon dioxide out of the atmosphere than can a normal agricultural crop and at the time of the laying down of the global fossilised fuel reserves in the Carboniferous Era, the world was largely covered in trees. There is some evidence[31] that one of the results of the Black Death in Europe in the Middle Ages not only resulted in a dramatic fall in population but that the abandonment of large areas of farmland resulted in reforestation and an increase in uptake of Carbon dioxide. This coincided with the start of a "mini Ice Age" although the workers at the University of Utrecht[31] conclude that there may be some doubt about taking a simplistic view of the correlation; it may have been more complicated than just trees and reduction of global warming.

Chapter 5
Relevant Technology
The Green Leaf and PCCSS
Composting as an Industrial Process

The Green Leaf and PCCS

Burning coal with new "clean" technology and CCS (Carbon Capture and Storage, such as capturing the Carbon dioxide which can be pumped deep underground into porous rock for storage) has been, intermittently, very much in the news as the energy debate progresses. However there is another route to bigger, cheaper and directly more productive CCS; agricultural land for biofuel production can deliver Carbon Capture and Storage in Soils. Indeed photosynthesis, a natural biological process of plants, can take enormous amounts of Carbon dioxide out of the atmosphere and pump Oxygen back in.

During crop growth, the chlorophyll in the green leaf of a plant traps the energy in sunlight. This gives the plant the energy to drive the chemical processes which take Carbon dioxide out of the air (via small holes in the leaf known as stomata) and water (via the roots) to form first sugars, followed by carbohydrates, oils and proteins. Indeed, during the Carboniferous Era, starting some 350 million years ago and lasting 60 million years, plants harvested sunlight, took Carbon dioxide out of the air and formed these compounds which eventually formed the black tarry substance which now drill for as "crude oil". Nowadays farming follows the exact same process, using crops to harvest the energy in sunlight. This is known as Photosynthetic Carbon Capture and the Storage is in the Soil[13][34][35]. The following figures show how this works.

Figure 5.1

The Basic Sustainable Crop Loop (the way it is often thought of and portrayed). The crop harvests sunlight and turns Carbon dioxide, taken in through the leaves, and water, taken in through the roots, into large Carbon molecules which we can use as fuel. If we burn the fuel, we release the Carbon dioxide again. What is often not thought about is that, in taking the Carbon dioxide out of the atmosphere, the plant actually gives us back the Oxygen. However, it is as well to remember that when we burn the fuel, we burn the same amount of Oxygen as originally released.

Figure 5.1
CROPS TO BIOFUELS - THE BASIC ROUTE

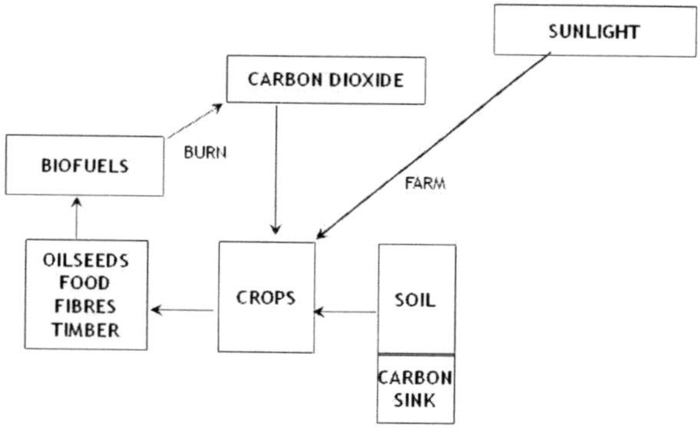

Composting as an Industrial Process

The micro-organisms in compost and land are different from us, of course. However, in terms of dietary needs, they are remarkably similar. They are different in that they can tackle much bigger molecules than we can. Just as we might put sugar (a 6-Carbon molecule) on our cereals for breakfast, the micro-organisms can tackle the lignin in wood (with an almost unlimited number of

Carbon atoms in the same molecule) and they never stop eating - 24/7.

As an example of this capability, it is interesting to look at one of Land Network's farms (a consortium of farms which recycle wastes to their own land). The example farm takes wood from local municipal and landscape gardening sources. Cellulose is made up of long chains of maybe 3000 monomers (groups of 5-Carbon atoms in a ring). So there are around 15,000 Carbon atoms in one of these chains. Lignin is made up of cellulose chains lying next to each other and joined by cross-linkages. This means that xylem (the "wood" behind the bark) in a tree may be almost just one molecule - everything is joined together. The number of Carbon atoms joined together is, obviously, enormous and almost so big that it is beyond ordinary number-comprehension. This farm also takes in PVA - Polyvinyl alcohol as a hot/warm liquid (it solidifies on cooling). As a liquid, it is comparatively easy to disperse it through a large mass of compost (preferably active and, therefore also hot). This is a big Carbon chain alcohol with many thousands of Carbon atoms in each chain. Even so, lignin, the basic molecule of wood, is even bigger with many cross-linked chains. As soil and composting micro-organisms can crack lignin, it follows that they can crack PVA. The BOD Biological Oxygen Demand (a measure of pollution potential) is enormous, so it is necessary to keep turning the compost but, as a result, the bioactivity rises rapidly and the process is faster. It helps make very good compost. PVA is a plastic; so we can and do compost plastic provided the format allows the micro-organisms to attack it.

So, compost heaps can break down almost any organic molecule - "organic" meaning molecules based on Carbon chains. Given a balance of other necessary food molecules, they will crack 100% of the Carbon chains in the feedstock in time. Furthermore, it does not really matter if there is any residue before spreading to land, provided that the soil is biologically active (meaning that it already has a reasonable amount of organic matter in it).

To say that the Carbon chains will be broken down is not the complete story. The micro-organisms make new Carbon chains in the form of hydrocarbons, carbohydrates and proteins. When the micro-organisms die, these molecules form a dirty black tarry substance (known as DBS) - initially very similar to crude oil. This product is actually what most of us call "humus"; it is the material which gives soil its black colour.

What composting can do is provide a "buffer" between a controlled process and the soil. That buffer can isolate physical, chemical and biological risks in order to allow processing, monitoring and safety controls to operate.

Composting is the standard technique for in-place remediation of heavy hydrocarbon contamination of soils[54]. If the process is sophisticated enough (as it is) to break long Carbon chains which are cross-linked and very stable (such as lignin in oak wood), then it can easily break comparatively short molecules that are found, for example, in petroleum or in a wide range of contaminants from industrial processes.

The only provisions to this are that the material must be spread out far enough and enough time given for the composting process to work. The compost process must have, of course, the basics available involving food, moisture, Oxygen and the micro-organisms. Mostly, "seeding" with a culture of "special" micro-organisms is not necessary; this earth is well enough populated with micro-organisms, wherever it might be.

Chapter 6
Closing the Loop
The Soil as a Processor
How the "Closed Loop" Really Works
Why Soil Universes Do Not Pollute Themselves

The Soil as a Processor

The processing capability of the soil dwarfs human industry. An American scientist once calculated that the micro-organisms in an acre of arable soil would weigh as much as a fully grown cow.

These are microscopic, tiny organisms of Nano-scale ("nano" means 10 to the minus 9, eg, one nanometre is one $100,000,000^{th}$ of a metre). Not just millions, not billions, but trillions of micro-organisms[9] in just a handful of soil, form a dynamic universe with enormous processing ability. Multiplication rates and biodiversity are enormous by human standards, and so is the range of their appetites. It is generally true that, given sufficient time, nature will deal with *any* material which has been spread out thinly enough, and bring the system back to a dynamic, balanced "normality" (where "normal" means sustainable).

How The "Closed Loop" Really Works And Why Natural Ecosystems Don't "Leak" Nutrients Or Self- Pollute

(Much of the material in this section on Nitrate pollution was first published in papers published in "Resource", Journal of the American Society of Agricultural and Biological Engineers in April 2001[14] and in "Landwards", Journal of the British Institution of Agricultural Engineers in Early Summer issue 2002[15] and is reproduced here by kind permission of those original publishers.)

Understanding the mechanisms in what is commonly called "the closed loop" and managing those mechanisms makes recycling to land dramatically safer in environmental terms. The figures show the principles of the closed loop. Organic materials, and inorganic ones which have food value for the soil micro-organisms, do NOT break down directly to form humus. Such materials added as "waste" are consumed by micro-organisms and turned into their own bodies. It is the breakdown of these bodies which form the stable black tarry material which gives soils their dark colour, generally termed "humus" [16] [17]. So, knowing how to feed these organisms is the first step in the management of composting wastes and of the soil. It is also important to see that the compost heap and the soil are not separate operations. Mostly, everything that goes on in a compost heap would also happen in the soil, even pathogen destruction. The big advantage to farming of composting before spreading to land is to use the temperature to kill weed seed. The micro-organisms feed, multiply and die, then break down into humus. Humus is an extremely complex mixture of heavy molecules of hydrocarbons (the same process which makes crude oil), carbohydrates and proteins (which lock up the Nitrogen). These molecules are large and insoluble and there is no limit to the quantities that can be put onto the soil safely. The evidence for this can be found in any natural ecosystem such as the Fens. When the Dutch engineer Vermoyden drained them nearly 300 years ago, some were 10 to 15m deep. Up until now, farmers could grow crops there every year, exporting the harvested products with the nutrients they contained, including the Nitrogen, and never need to add any fertiliser. Clearly, there had been an enormous reservoir of crop nutrients but the Norfolk Broads are not polluted with green slime and dead fish. Not all Nitrogen is the same! NVZ's are "Nitrate Vulnerable Zones" and are the UK government's attempt to create regulations which will prevent Nitrate pollution of groundwater. Incidentally, the evidence is that it does not work.

Those large molecules of organic matter as humus will remain forever until long strands of soil fungi, called mycorrhiza, linked at one end to plants requiring food, start consuming them. These mycorrhiza either go up to and envelop the plant root hair, rather like the placenta in a baby mammal in the womb, or actually cross the root hair wall into the plant. As common sense might indicate, the plant and the fungi evolved together over millions of years and they operate at the same soil temperatures, so the system is demand-led. This system is how all natural ecosystems not only eliminate Nitrate pollution, they eliminate all such out of balance pollution including phosphates, potash etc, etc. There is a further advantage, as the system locks up Carbon in the soil. The 100 million tonnes of "waste" produced in the UK per annum and which could be recycled to land would, if incinerated, produce around 75 million tonnes of Carbon dioxide per annum which is 10% of the Kyoto Protocol estimate of total UK emissions. Composting to land can lock that up.

The figures on the following pages show how natural ecosystems manage to "leak" enough, and only enough, to keep the system working without pollution or starvation, i.e. in balance or "sustainably". Figure 6.1 shows a conventional view of how the system works.

Figure 6.1

The conventional view of how plants feed with the assumption that nutrients get to the plant via solution in the groundwater. With mineral fertilisers, this is probably either partly or completely true(13).

Figure 6.1
HOW THE PLANT FEEDS - MINERAL FERTILISER

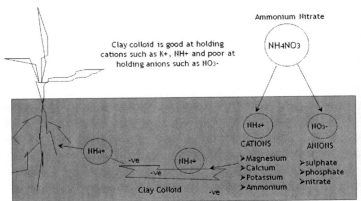

When mineral fertilisers such as Ammonium nitrate are applied, the cations are held in the soil colloid "bank" which also holds water. However, rain will take nearly half of the nitrate into groundwater.

Although not incorrect, especially when referring to agriculture which uses mineral fertiliser, this description is incomplete and potentially misleading, especially if the concept of this loop were applied to soils with significant levels of humus or organic matter. Further, it does not explain why natural ecosystems don't leak enough to cause pollution. Figure 6.2 shows how such pollution is avoided and shows the mycorrhizal conduit which is the central mechanism in what is commonly referred to as the "Closed Loop". It is the mechanism which stops leakage at a level of pollution. It is this very same mechanism which feeds plants (see (18) (19) (20) (21)) and protects them from disease (see also (22) (23)).

Figure.6.2

In natural ecosystems, plant nutrients do not enter into solution in the ground water in order to enter the plant. Humus is a complex mixture of heavy molecules which are not soluble in water. Neal Kinsey, in his book "Hands-on Agronomy"[20], points out that this humus has several times the colloidal capacity of clay and will hold onto anions as well as cations. That, however, still did not explain how the nutrients got into the plant without leakage. It was the American PGA (Professional Golfers Association) who pursued this investigation to show that the soil fungi, known as mycorrhiza, fed at one end of their hyphae on the humus and the other end went not up to somewhere near the plant root hair but actually cross the root hair wall into the plant. This finding was added to by researchers at Aberystwyth in South Wales who showed that there was another type of mycorrhiza which went up to the root hair and wrapped around it much as the placenta in a mammal. This is a molecular level relationship and a closed conduit. That is why the natural ecosystems do not leak.

Figure 6.2
HOW THE PLANT FEEDS - NATURAL ECO-SYSTEM

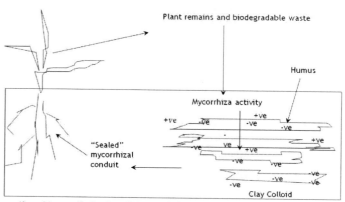

Mycorrhiza are the key to pollution control because they give a "Closed Loop" to recycling both cations <u>and</u> anions.

49

Figure 6.3
What composting can do is provide a "buffer" between a controlled process and the soil. That buffer can isolate physical, chemical and biological risks in order to allow processing, monitoring and safety controls to operate[1].

Figure 6.3
HOW THE PLANT FEEDS - RECYCLING WASTE

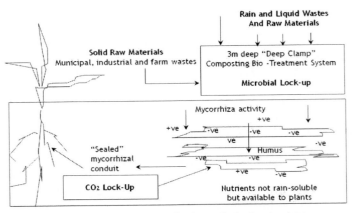

How the closed loop gives pollution control and scope for treatment systems.

Why Soil Universes Do Not Pollute Themselves

Soils manage pollution in two ways. It is, perhaps, useful in understanding how these mechanisms work to first understand what pollution actually is. Pollution is, incidentally, not just quite natural but fundamental to life itself. *Part of* the definition of a living organism (as distinct from not living – a tractor or a computer for example) is that the living organism is producing pollutants. These pollutants are products from the body of the organism which must be got rid of, outside that body. The question then arises as to when that production of pollutants becomes "pollution". The answer, in the scientific sense, is *always*. In the practical or legal sense, the exact definition of pollution depends on the following. If the production of pollutants

50

is at a level where the local environment cannot, in time, bring the system back to what the ecosystem previously was, if there is a shift in the ecological equilibrium, then, it may be said, pollution may have occurred.

The two ways a soil combats pollution are by providing a "buffer" to buy time and by digesting the pollutants and passing them into the ecological chain.

Soils which are substantially sands have little buffering capacity and little ability to hold chemically onto any particles – large or small. For example, by adding ammonium nitrate fertiliser to sand, the ammonium cations and the nitrate anions will leach out very easily with probably more than half going into the groundwater with rain or irrigation. That is a significant economic loss and potential pollution of groundwater. However, add the same material to a clay soil and the colloidal capacity of the clay will retain much of the ammonium cations and possibly some of the nitrate anions, too. Add the same material to humus and there will be a retention of all of both ions[20]. There will be no leaching with rain or irrigation of either the ammonium or the nitrate ion and pollution of groundwater will be eliminated[9] [13]. So, different soils will have different buffering effects and we can alter that capacity by adding and managing the organic matter levels of soils, specifically the humus content.

An interesting trial was conducted by Rose *et al*[49] in a contract commissioned by Defra on the problems of spills or washings from field crop sprayers. These washings would be expected to contain very active ingredients which, in any concentrated amount released in an uncontrolled way, could cause pollution. The trial involved adding a cocktail of commonly used spray chemical active ingredients onto a series of filter traps such as gravel or sand. One of these traps was half a metre deep with straw, "non-peat" compost and loam. The last trap, they called it a "bio-bed", was almost totally effective in breaking down the substances and eliminating pollution risk. That bio-bed was half a metre deep and worked at ambient temperatures which, in the UK, would rarely be above 20 to 25°C. In Deep Clamp composting, Land Network[1] uses a "bio-bed" 3 metres deep and maintains temperature ranges of 50 to 90°C.

So, it is clear that soils, and even more so, compost heaps, can and do have very sophisticated and capable antipollution capabilities.

This leads us to a conclusion and, of course, naturally, to a further question. Firstly, yes, it is possible to manage these mechanisms by introducing a buffer (such as a compost heap in a controlled and "imprisoned" situation) and by providing conditions which will encourage and manage the biological activity, thus destroying pathogens and undesirable toxic molecules. The question is how to identify and manage the need for sufficient time and dilution to allow these desirable functions to reach an identifiable and acceptable end point within a predictable time frame. The answer to this is known as "Dispersion Technology".

Chapter 7
Making it Happen
The Saving Grace - Proximity Principle Recycling of Wastes
How to Manage Solid Wastes by Composting, Feeding and Direct-to-Land Operations
- Alternative Methods and Technologies
- Shredding and Screening
- Deep Clamp Composting
- Deep Active Bedding of Stock
- Applying Liquids to Compost
- Comparison of Systems - Mechanisation
- Spreading to Land
- Direct Incorporation
How to Manage Liquid Wastes
How to Manage Biosolids
How to Find Suitable Wastes
How to Manage the Nutrients in Waste
Process Capability and Safety

How to Manage Solid Wastes by Composting, Feeding and Direct-to-Land Operations
John Lawes, generally regarded as the "father" of the mineral fertiliser industry, did not start his first Superphosphate factory until 1843. Progress was initially slow but "artificial" fertilisers eventually became cheap to produce, very concentrated (simplifying transport and spreading), and they made dramatic and visual differences to crop appearance and yields. Once large scale production became possible, it took only 15 to 20 years to sweep the world. These artificial, or mineral, fertilisers are a comparatively recent invention. Until they arrived, farmers used mainly locally available wastes, but also guano (bird and bat "droppings" or faeces), transported across the oceans by the great sailing ships, for this same purpose of fertilising crops.

When locally available wastes, such as cotton "shoddy", leather trimmings, slaughterhouse wastes, were available, they were used as fertiliser. Grain was imported from overseas and fed to livestock, and the urine, dung and soiled bedding (muck) was used for the same purpose. So was sewage. Waste has always been recycled to land, ever since *Homo sapiens* began to wander and even before he settled. It is important to understand that everything originally comes from the land and it will eventually go back to this same land. Given sufficient dispersal rates and time, nature will have its way.

The basic process of composting, or any other aerobic biological process, involves four needs; a feedstock with reasonably balanced nutrients, an appropriate population of micro-organism species, moisture (water), and Oxygen.

Generally speaking, if the material is of plant or animal tissues, it will have enough nutrients to react reasonably naturally in a compost process. In most environments, there are plenty of micro-organisms and of a wide enough range of species to operate the process. If there is not enough Oxygen, the process will go anaerobic, become odorous and slow down. If there is not enough water, the material may well go dark in colour but the process will be incomplete and will start up again when the material is incorporated into the soil after spreading[8][9]. This may not matter unless there is a shortage of Nitrogen and, in such a case, the micro-organisms in the soil may preferentially use the easily available Nitrogen in the soil to build their own bodies in order to attack the energy source (the Carbon) in the added compost. Farmers sometimes call this "Nitrogen starvation". The Nitrogen is not lost and will be available to the crop at a later date when the biological activity of the soil catches up with balancing the Nitrogen status of different fractions of the soil. This was described in a composite form in "The Straw Manual"[21]. Previous to this date, farmers in the UK faced a ban on burning straw behind the harvester and it was necessary to find out how farmers could incorporate unwanted straw into the soil without high costs and/or loss of yields. That book collated the available research from world-wide sources and showed that the soil will live quite well with what might appear to be enormous imbalances (as with the high Carbon content of cereal straw) provided it is given a little help and time. The basic rule when changing to a system which involves putting large amounts of Carbon into a soil is to add enough Nitrogen fertiliser in the first year to allow the soil micro-organisms to build the protein of their own bodies, allow a little less Nitrogen in the second year and less or none by the third or fourth year. After that the soil system will cope because the added Nitrogen does not leach out and the soil micro-organism population has changed in species and population numbers to give the biological activity required to deal with the new regime.

A compost process is basically the same. It is worth going back to the basics of composting for a moment; the four needs of feedstock, micro-organisms, moisture and Oxygen. It might help to add "time" at this point. If one of these basics is either not there, or not in the right quantity, then the process will change or slow down. For example, if there is not enough air, the process will slow down and either stop or go anaerobic which will give off a bad smell. If there is no change of gases, the process will eventually stop. Similarly for water; lack of it will slow the process eventually to the point of cessation.

Exactly the same applies to the other two inputs of feedstock (obviously) and micro-organisms (less obvious but quite interesting).

An idealistic view of a composting process graph of temperature against time is shown below. In practice, this will only occur in a compost of uniform material, uniformly shredded and uniformly aerated. Most actual operations will produce a more patchy picture although of the same progression.

Figure 7.1

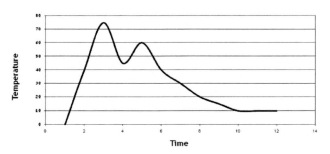

Figure 7.1
COMPOST TEMPERATURE CURVE

The first peak in temperature is mainly of bacterial activity. This is the risky time in terms of lack of Oxygen and odour production. Deliberate Oxygen starvation can and does produce Methane gas which can be used in heat and power generation. The top of this curve will normally be targeted at 55 to 75°C. Above this range, the temperature gives rise to increasing losses of Nitrogen (which is, of course, valuable as a fertiliser) and increasing risk of fire. The second peak is of fungal activity; mainly "Pin Moulds", which are Penicillins. The trough in the middle is mainly of Actinomycetes, the fungi which give woodland its pleasant smell after rainfall. The curve on the right falls, but never actually to ambient, unless the material, or the ground onto which the compost is spread, is frozen solid.

Following the logic of that curve and the four basic inputs into the process, it becomes clear that it is possible to influence the direction and speed of the process. It is possible to speed the process up and to slow it down but never make it fast. It is possible to make the process more odorous or less odorous but never completely odourless. It is possible to make the output look dark in colour and friable, but that does not necessarily mean that the process is complete and all the soluble nutrients have been incorporated into the bodies of micro-organisms, turned into humus and made safe to spread without pollution risk. Thoroughness can be speeded up but never made fast in composting. This is a biological process and time is fundamental to that process.

Alternative Methods and Technologies
If differentiating by detail, there are almost as many methods of composting as there are operators. However, there are two fundamentally different approaches – and, of course, every variation in-between.

Windrow composting, strictly speaking, is composting in long thin rows of material. They vary significantly but may be as small in cross section as 1m deep and 3m wide, turned with a multi-tined rotor machine that straddles the row. The other distinct type is known as static pile, Table-top windrows or Deep Clamp Composting. This is a large heap, around 3 to 4m deep and is turned with some sort of mechanical shovel or 360° excavator bucket.

The Birth of Land Network

Land Network is a network of farmers who recycle wastes to land. They pool knowledge of process, regulation, commercial data and everything and anything to do with recycling-to-land sustainability. It was developed under the Enterprise Initiative of the DTI back in the early and mid 1990's. (The DTI is the Department for Trade and Industry of the UK Government.) The organisation is described in more detail in Chapter 10 and further details can be found at www.landnetwork.co.uk.

When Land Network was first conceived under the original DTI programme, there was a need to choose a method of operation. The textbook systems were examined, especially the then fashionable windrow system with straddle type windrow turner. Most of the methods discussed in the technical media were of imported technology, using major pieces of equipment and with significant advertising budgets. One by one, these were discarded, not because they did not work (they all did – at least in some sort of way), but because they did not fit the criteria of using what already existed in the rural economy and, most importantly, did not fit the skills and culture that managed the landbank, i.e. farmers[26].

At the time of having to make a choice, back in the early 1990's, it was necessary to look at the classic windrow technique using a straddle-type turning machine. Even then, the price of such a machine would have been over £100,000. The windrows would have been, typically, maybe 1m deep in the centre and 3m wide. When it rained, the windrow would thatch in heavy rain and the run-off would run in a stream down each side. There would be a question of mixing the edges to the central heated area. Of major concern was, and still is, the fact that the rotor was designed to mix air with all of the windrow. The centre of that windrow would contain bacteria (some of which might still be pathogenic in the early stages of composting), fungi and volatile chemicals. These would be raised into the surrounding atmosphere. Indeed, when one of these machines is working, there is a cloud of steam around the machine. That steam would be laden with those organisms listed above and now often referred to as "bioaerosols". There was a perception of health risk. But what really directed a look elsewhere was cost and failure to use what already existed on the ground.

For those of us at the time who had grown up in the agricultural industry, it was common knowledge that farmers had, for centuries, been making large heaps of "wastes" in a corner of their fields. Indeed, it was one of only two ways of making fertiliser for the land. One was to use green manure crops including legumes which could fix their own Nitrogen. The second was to use "wastes". Most people, even of urban background, knew that farmers used the manure from livestock to fertilise land. If the stock had been fed on food imported onto the farm, such as grain, (as opposed to home grown food) then the nutrients available to put to land would increase. What few people in the UK can now remember is that farmers also routinely took local wastes and put them to land usually via a large storage heap. The storage heap would be slowly built up and it would stay there maybe 2 or 3 years, sometimes much longer, and wait for rotting down and a suitable crop window. They did not usually call it "composting" but that is what it was. Interestingly, they usually did not turn it at all. The material was not shredded either. It was a slow process and it did not usually smell. Surprising to many today, this was a common, normal practice, without odour, nuisance or pollution.

The design of the heap was very important. The shape was clearly to be flat-topped to absorb the rain, necessary as moisture in the process and to stop run-off. But what depth should the heap be? At the time, it was reckoned that a small garden compost heap needed to be around 1.5m deep to avoid rain running through. So, arbitrarily, that figure was doubled to 3m. Later it emerged that this allowed the best compromise of letting the heat out and the air in.

Figure 7.2

Figure 7.2
HOW DEEP CLAMP COMPOSTING WORKS

OBJECTIVES
1. To aerate the clamp, maintain moisture and retain heat up to 60°C.
2. "Boil off" surplus water.
3. To thoroughly pasteurise the products

Turning   Additions
Dumping and Turning Area

RISK CONTROLS
1. Start with straw or similar to aerate mixture of waste.
2. Make clamp 3m deep from municipal, industrial and farm wastes.
3. Manage turning to let air in and control temperature in the 50 to 90°C range.
   Note: 55-60°C is ideal to control pathogens.
       60°C is often regarded legally as the ideal for disease control.
       50-90°C is the range which is, in practice, the likely operational range.

The next diagram, Table 7.3, compares two composting methods (not to the same scale). On the left, windrow composting. On the right, a large heap with enough depth to absorb rainfall and hold it. If the heap were vertical sided and the rain vertical, then the rain outside the heap had fallen there for millions of years without

60

harm to the environment. In practice, the sides of the heap are not vertical, nor the rain, so there is a small edge effect which the soil can easily deal with.

So, the research and thinking at the time moved away from the windrow system. This was partly because, at the time, time in the process was not important. The research conclusions were summarised at the time as in the following table.

Table 7.3
Table: Comparison of Windrow composting with Deep Clamp Operation
Source: DTI research contracts carried out by Land Network International Ltd between 1992 and 1996[20].

This research programme looked at the deficiencies of centralised composting in general and windrowing in particular. The conclusion was to develop existing garden and farm composting techniques and work on de-centralised, rurally-based systems which:

- recognised where the market was i.e. farming then bought around £900,000,000 worth of fertilisers pa and needed the organic matter;
- recognised where the resources already existed space, machinery, labour, skills; and
- recognised that farm-based systems are basically responsible whereas centralised systems with pressurised, target-meeting, absentee managers and go-home-at-5pm operators are basically not responsible.

WINDROW	DEEP CLAMP
Operating Area Commercial input example of 1000t would take approaching 1ha to set up. This is normally on concrete.	As little as 25% even down to 10% of the area. Often not on concrete, just hardcore.
Rainfall & run-off In heavy rainfall, for example above, there would be 2400 meters	In a well managed clamp, total elimination of run-off and leachate. The operation

of dirty brown water running down the sides of the windrows.	actually gets short of water and may need to import liquids for best results.
Pasteurisation Poor surface to volume ratio with approximately 8 sq m of surface area per tonne and shallow depth of insulation. Therefore poor heat retention, especially at the edges and poor pasteurisation.	About 0.18 sq m of surface area per tonne and 3m depth. Easy to control and continuous good pasteurisation. At least 6 log reduction in pathogens (one millionth) and maybe 7 log (one ten millionth).
Chipping Necessary either by pre-chipping or by the rotary windrow turner.	Not necessary but some sort of shredding is normal to speed up the process.
Process time Financially important. Usually 4 to 12 weeks.	Financially not important. Not usually less than windrowing. If thick material not chipped, then much longer.
Odour Probably low odour but some materials difficult.	Less odour. Maybe very little odour with most materials. Can manage well to give low odour on very difficult materials.
Mechanisation Normally requires large, expensive and not-road-mobile machinery.	Normally operated with existing farm machinery.
Centralised Because of the machinery and concrete, normally centralised. Therefore major operation and probable NIMBY opposition. (Not in my back yard.)	Normally de-centralised with 65% to 85% less transport distances. Small scale farm operation with a degree of isolation. NIMBY very rare.
End Product Normally needs shipment elsewhere.	Normally used on site. Possibly a contribution of up to £900 millions to UK balance of payments.

WINDROW	DEEP CLAMP
Production of aerosols and airborne fungal spores. A major problem at turning. Probable risk to health. Future regulation may necessitate building cover and air filtration.	Possibly a 7-log reduction (i.e. one ten millionth).

Shredding and Screening
The problem with mechanical treatment such as shredding is that it requires a lot of energy and, therefore, has a cost both financially and environmentally. Big shredders may consume 50 litres of diesel fuel per hour. The advantage is that smaller particles will be easier to handle mechanically and the process will be faster. However, green waste from twin-bin municipal collections very often will not really need shredding; particularly if the final product is to be put to heavy soils where large residual particle size is an advantage. So, the garden compost as a fine, uniform, black "crumb" is not what is required in most arable farming operations. This implies that screening, to remove the larger bits, may not be needed. Indeed, most of the farms in the national farmers' consortium, Land Network, so far, do not screen at all, or only selectively. Screening, however, may be an alternative to shredding. For example, with suitable twin bin green wastes (which may be of low branch size but are often heavily contaminated with plastic and other household wastes) it may be better on all counts to put the whole lot into the composting process without shredding. Screening out at the end will make the plastic much easier to pick and the oversize particles can just go round again.

Deep Clamp Composting

There are many labels which get attached to large heaps of waste undergoing the compost process. The simplest is "a compost heap", but there are many that sound more technical, including "static pile", "table top windrows" and "Deep Clamp".

OUTLINE OF THE PHYSICAL PROCESS
DEEP CLAMP COMPOSTING

1. Only source-separated material of organic nature is used.
2. It is delivered to a farm site, usually off concrete, with no building of any sort, and stacked in a heap 3m deep. The reason for no concrete is that concrete collects rain and allows it to drain into the *bottom* of the heap. Rain going in at the *top* of the heap will steam off because of the heat generated in the heap. The heap is fairly tidy with near vertical sides. This 3m depth was originally based on the gut feel of experience. There has been some significant research[8][9] on this potentially critical aspect of compost process temperature since. This confirmed the functional desirability of a deep heap in the on-farm situation. A deep heap can and will remain aerobic provided the shredding of the material is relatively coarse. There is a further characteristic of the deep heap and one which initially appears to be a disadvantage as it often means the process is not quite as fast as windrow composting and is often a little cooled. The advantage was shown later with research on best anti-biotic activity, carried out at the University of Hull, which showed that 55 to 60°C, for a longer period than processing at 60° plus, gave better pathogen kill.
3. In the UK, this recycling of waste to land could, at the time of writing, only legally be operated with an

Exemption from the Environmental Permitting Regulations registered with the Environment Agency, or with a full scale, much more involved, full Licence or Permit. Again at the time of writing, for the Exemption, amongst other restrictions, this entailed operations with not more than 1,000cu m in the process at any one time. This, at 3m deep, will be approx 18m square, i.e. a little less than a cricket pitch down each side. This UK regulation has been under review for some time and while new regulations have been earmarked to come into force on various dates and, up to the date of publication, repeatedly postponed, there will, eventually, be changes. The process can legally be carried out in much larger masses under full Permit conditions. Whatever the regulations, there is no technical limit to the size of the operation.

4. There is some significant technology in the rules of how the heap is put together and managed so as to prevent leachate and prevent anaerobic conditions. This can be controlled in practice with a relevant Code of Practice.

5. 100% of the material in the Land Network programme is spread to land and incorporated (ploughed or cultivated to mix in with the field soil) when there is a crop window. It may also be spread behind a silage cut. Finely screened composts can be put to growing crops, including grass. The material is almost completely odourless on maturity and spreading.

Managing Odour Through the Temperature Curve

Look again at the basic temperature curve in the compost process, repeated below. The basic temperature curve is never, in practice, as tidy as is shown in the graph. The main risk of really bad odour is during the first peak - the bacterial phase. High activity and low Oxygen supply mean the process is likely to be anaerobic and bacterial and offensively odorous. Going anaerobic in the trough is not likely to have serious risk of bad odour and going anaerobic in the second peak and onwards will certainly make any "musty" smell of the fungi worse - but unlikely to be as offensive to the human nose as anaerobic activity in the first, bacterial phase.

Figure 7.4

Figure 7.4
BASIC TEMPERATURE CURVE

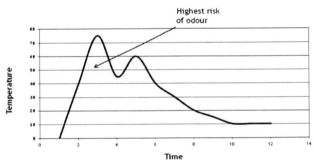

In reality, some materials will process faster than others. The more rapid the process, the more likely it is to be relatively uniform and at a particular stage. Therefore, in very rapid situations, such as with catering wastes and with rotary turning every couple of days, the risk of odour in the first stage is at its highest. Conversely, in slow situations such as with woody green wastes, not turned very much in Deep Clamp processing; only some of the material will be active, and probably not very active, and, therefore, the risk of odour is the lowest.

The simplistic conclusion on odour control is that the highest risk is in the first peak phase; anaerobic bacterial action here is likely to be offensive. There will always be some mustiness in the second peak and onwards and this will be made worse by lack of turning if the mass of the compost is active. There is a way of reducing the most offensive risk. Consider the following variation.

Figure 7.5

Figure 7.5
NORMAL AND LOW TEMPERATURE PROCESS CURVES

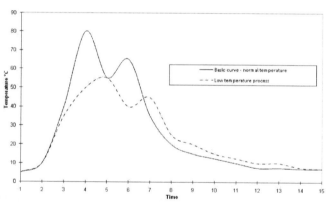

The second curve above shows a lower first peak temperature. If temperature can be limited to less than around 65°C, for example by not having quite enough of a particular food which the micro-organisms need (commonly, this is Nitrogen) and by not reducing particle size too far, then the rate of micro-organism activity will reduce and the risk of odour in this phase will also reduce. This can be easily managed with Deep Clamp, or "static pile" composting; it is slower, but safer.

In practice, one of the advantages of Deep Clamp Composting is that it can be slower and because it has a very low surface to volume ratio, the scope for odour is also very low. If it is run with large particle size (so that it can breathe), of not very high Nitrogen materials, and with little turning in the first phase, then there is not likely to be significant odour of the offensive, anaerobic bacteria type which can, sometimes, emit from that first period of the process. However, the rest of the process, as always, is likely to have a musty odour, characteristic of fungi. Again, running a slow process, with little turning, will minimise emissions.

In practice, there is a compromise between turning to keep the process aerobic, slowing it down by not turning to reduce emissions, and the speed which is required for commercial or regulatory reasons.

Deep Active Bedding of Stock and Outdoor Corrals

It is technically and practically quite attractive to build up a deep pile of organic material and let animals, particulartly pigs, do the turning. The plan here is to use a fence to retain the animals, possibly and usefully a wall of large bales of straw, and put a layer of "waste" on the ground, followed by a period of "grazing" by the stock. Sheep and cattle will do this but keep the layers thin and add new material little and often. The layers can be thicker and of less critical periods between additions by using pigs which will root through the material with their snouts and aerate the mass to maybe up to half a metre deep.

If the "waste" has significant food value to the animals concerned, it may be possible to avoid adding food supplementation, the animals living entirely on the waste. A variation on this is to run "barley" beef cattle on the deep bed and run pigs underneath them. The pigs will find enough to eat (including of the cow dung) without supplementation. With care in stock density, this can work quite well.

A further variation is outdoor stock "corrals" which are based on the idea of a high fence around a well-drained base, with layer upon layer of bedding added. The "bedding", in this case, being wastes with the sole purpose of bedding rather than having some food value. The advantage of this system is that it does not involve the expense of a roofed building. The disadvantage of the system is that, in times of rain or snow, it involves significant added bedding – more than if the stock were housed under cover. That is significant if the bedding has a cost. However, if the bedding is "waste" and has an attached income, then the corral becomes a processing factory to turn the waste into farmyard manure. This can be an added facilitiy for recycling to land via a logical, safe and profitable route.

Figure 7.6

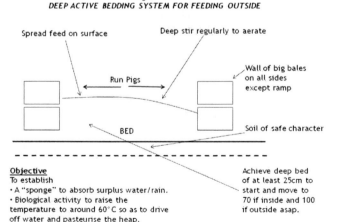

Figure 7.6
DEEP ACTIVE BEDDING SYSTEM FOR FEEDING OUTSIDE

Spread feed on surface

Deep stir regularly to aerate

Run Pigs

Wall of big bales
on all sides
except ramp

BED

Soil of safe character

Objective
To establish
• A "sponge" to absorb surplus water/rain.
• Biological activity to raise the temperature to around 60°C so as to drive off water and pasteurise the heap.

Achieve deep bed
of at least 25cm to
start and move to
70 if inside and 100
if outside asap.

69

Applying Liquids to Compost

Generally speaking, making composts with solid materials but without the addition of some liquids does not work very well. Of course, if the materials are fairly wet at the start, it can be managed. However, in such circumstances, anaerobic conditions and the resultant odour is a greater risk and there will certainly be a need for careful monitoring and probably more turning.

Addition of liquids may be by spraying, preferably piping the liquid onto the top of the heap with a dribble bar (to avoid spraying liquid up in the air), by "lagooning" on top of the heap or by what is best described as "dunking". The advantage of dribbling onto the top of the heap is one of control but it may, under some circumstances, mean that a person has to climb on top of the heap – and that is best avoided; the pipe is best positioned and secured using a telescopic loader. The lagooning method involves a compost heap of at least 3m depth and the use of a crowded loader bucket to create a dry lagoon on the top of the heap. The lagoon needs be maybe half to three quarters of the volume of the delivery tanker (depending on the viscosity of the liquid and the absorptive capacity of the compost in the heap). The tanker next pumps its load into the lagoon and this, with a 25 tonne typical load and truck pump, will take 20 to 30 minutes. The process is monitored for leakage but usually allowed some absorptive time. If there is leakage or after 20 to 30 minutes, the loader bucket is used to mix the two materials rather as in mixing concrete by hand with a shovel. It works quickly, safely and at low cost.

An alternative, the "dunking" method, is low cost and safe, ideal where a site has a run-off lagoon or trough in which a loader can work. In Land Network sites, such a trough, found at the end or at placed about one third of the length of the compost area slab is common. If liquids are pumped or dumped into that trough, then new and shredded solid materials for composting can, at delivery, be dropped into the trough to mop up the liquid. The wet solids can then be added to the compost in the normal way, allowing run-off to go back to the trough. Alternatively, if the trough is designed to be just a little wider than a loader bucket, the liquid can be scooped up and bucketed onto the compost heap. A further alternative is to pump the material and spray it onto the heap.

Where a tanker can be equipped with a large diameter hose which can be dragged over a large heap with a loader or 360 machine, then very large amounts of liquids can be added and this was one of the reasons Land Network chose the Deep Clamp method of composting. With around 3m of absorbent material below the hose, this is the best possible structure/least risk which will retain liquid in the compost mass, and the easiest and lowest cost for mixing.

It is worth stressing three things with regard to adding liquids to compost. Firstly, the process is a biological process which depends on moisture. Secondly, the right amount of moisture is critical to the micro-organisms consuming the solubles and "locking them up" into humus and making subsequent spreading safe. Thirdly, if the moisture is added as a "waste", then it is recycling and financially rewarding.

High BOD Materials such as Bioglycerol Oil/Fats
Fatty and oily materials are often available for composting as liquids. Some will be solids at ambient temperatures. The good news is that such materials are

almost always based on large Carbon chain molecules thereby presenting a lot of available energy, with the result that they can really activate composting process. These materials often command high gate fees as such a high level of activity is necessarily difficult to manage. Generally, the first problem is, as the designation of "high BOD" implies, high Biological Oxygen Demand. In practical terms this means that the heap will need frequent turning in order to avoid anaerobic and odorous conditions. Managing this needs a competent operator with some "feel" for the composting process. Temperature is a useful guide and it may help in large operations or if skills are still developing, to use an Oxygen metre. As a guide, if the temperature stays up to normal levels, not less than 50°C, and does not dip at all, the process is still aerobic. Regular turning in line with temperature, will maintain this aerobic condition.

Initiating the active process may be difficult and there are several factors influencing that start-up and the continuance of process. The nature of the absorbing solid material is critical. It needs to be only fairly coarsely shredded, in order to hold air. Somewhat in conflict with that is the nature of the viscosity of the added liquid. Some liquids may only be pumpable at well above ambient temperatures. They become more viscous on cooling, sometimes to the point of solidifying. Examples of this sort of material include PVA (polyvinyl alcohol) and, at low ambient temperatures, bioglycerol (from biodiesel production from virgin crop oil). To hold these materials in the compost heap is easy enough at the start as the material cools. However, if only a small amount of material is added to a large mass of active, very hot compost, then the liquid stays liquid and may simply run through material if it is too coarse. This situation can be very much worse if a large amount of liquid, especially bioglycerol, is added to less active, cooler, compost-in-progress. When the process re-activates, sometimes in two or three days time, the temperature inside the heap warms up the material again and it becomes less viscous and may run out of the bottom of the heap again. Getting it right is largely a matter of experience in adjusting the initial quantity and its dispersal in the available solids, as well as in monitoring and reacting to developing conditions. The bottom line is frequency of turning and a simple water/leachate collection system. Generally, if material of the type which does solidify at ambient temperatures does leak out of the bottom of the heap, it will solidify quickly and not go into the liquid collection system; it can be swept/scraped up by mixing with compost. If it does not solidify, it will get carried into the collection system - another reason for designing a simple, open-to-the air, easy-to-pump or scoop-out system.

Comparison of systems - Mechanisation
Crop Windows, Sacrifice Land, Reclamation of Land for Energy Crop Production
In tropical areas, man grew up with a "slash and burn" operation in the forest. Trees were cut down, burned and the ash sustained crops for a few years and then the people moved on and repeated the process on another spot. In some areas, the same practice is still in use today. The reason for moving on is that the soils, on their own, are often not very fertile and the rainfall such that nutrients are easily washed away into the groundwater. The fertility is locked up in trees, plants and the detritus (organic wastes from falling vegetation) on the forest floor.

Comparably, intensive farming in developed countries has exploited the land in a similar way and with similar consequences. Use of high-powered cultivation tools and high inputs of soluble mineral fertilisers have allowed organic matter levels to fall. Cultivation assists the soil micro-organisms to oxidise the large Carbon molecules to produce Carbon dioxide. This reduction in the large Carbon molecules results in changes in the physical strength of the soil structure. As that happens, there is a need for higher power inputs to cultivate the soil; there is a spiral of inputs and the soil is only kept productive by increasing inputs - all of which are energy demanding[27].

As discussed in Chapter 4, reducing cultivation by using "Minimal Cultivation" or what is termed "Zero Till" in the USA or "Direct Drilling" in the UK, will slow down this oxidation of organic matter[7]. Research by Land Network International[10] for ICI Plant Protection in the late 1980's/early 90's, indicated that oxidation of the organic matter occurred from something in the region of 35% per annum on a declining basis with conventional high-input cultivations to around less than 10% with direct drilling. "Declining basis" is comparable to "half life" with radio activity decline. If the figure for Carbon release is 35% this year, it will be 35% of what's left next year if the cultivation regime remains the same, and so on.

As all developed societies produce "wastes", there is a logic in composting what is useful and safe and putting it onto the land to replenish organic matter levels. There is here, however, a potential problem. If the ground is not productively used for crop growth, there is economic loss (the area is not producing anything) and the ground is at higher risk or erosion from wind and rainfall. If it is used for crop growth, it is certainly a potential difficulty to apply compost or wastes to that ground. Composts and wastes tend to be high volume/low value and, therefore, the tonnage needed per hectare may be enough to smother or reduce the growth of the crop. So, there needs to be care in these applications, either in appropriate preparation and application rates or by spreading in "crop windows", i.e. between one crop harvest and the planting of the next.

Sacrifice Land
If the compulsion to spread to land is high enough (by physical necessity, regulation or by financial incentive), then it may be necessary to "sacrifice" land by spreading over a growing crop. Historically, to most farmers, this would be little less than sacrilege. However, as recycling wastes becomes more compelling (again by cost/absence of mineral fertiliser or by financial incentive from wastes), and/or where there may be depressed crop values in some parts of the developed world following over-production, spreading may become more important than crop production.

Reclamation of Land for Energy Crop Production
There is, of course, an area of land which does not have the complication of existing crop; that of reclamation. Reclamation may be from previous industrial activity or from desert or soil-depleted or soil-less upland (although that, too, is quite likely to be from previous human activity)[12].

That, from a recycling waste point of view, is clearly an attractive option. However, as later chapters in this book will show, there is also a compelling attraction of this in terms of environmental stability and the very survival of the human race[1]. We really do need to look at the reforestation of this land, preferably for trees and crops which are energy-producing.

75

Spreading to Land

Generally speaking, if any material is spread out far enough and enough time is given, soils will deal with anything, however "toxic"; this is known as "dispersion technology". In farming, they have been spreading muck for centuries. Farmers have an instinctive understanding of what is good for their land. It is in the nature of regulators that they are unlikely to understand this and, therefore, will not easily trust. However, regulators do have a point. While the farmer may be trusted up to a point, and that local, in-depth experience is never anything less than invaluable, it is also necessary to recognise that wastes may not be what they used to be and that the regulator has a responsibility, too. Despite that recognition, there is a difficulty in that most regulators are not hands-on agricultural scientists. It is inevitably true that responsibility without adequate technological training for the job will result in inhibition and slow down in activity and serious difficulty for the innovative or entrepreneurial.

Cation Exchange and Base Saturation

Back in the1950's, Dr William Albrecht[28] began to look at soils which had been, within living memory, broken out of prairie. These soils had initially yielded crops well but had gradually declined. Whatever farmers did in terms of addition of fertilisers, they could not get back to where they started in terms of crop yield. Albrecht developed chemical models of the prairie soils and the depleted soils and then investigated practical ways of applying available materials to move the depleted soil back to prairie status by using the models as a guide. It worked and still does with Neal Kinsey widening its scope and application[20].

Think of feeding the soil and the soil feeding the plant. Now think of the soil as equivalent to a cow's rumen with the alimentary tract as the soil micro-organisms. The soil has enormous numbers of micro-organisms. Neal Kinsey[20], soils expert in the USA, reports estimates of the number of micro-organisms in one acre of soil are equivalent in weight to one whole cow. Maybe, according to some thinking, the soil mycorrhiza are the last link in the chain leading into the plant root hair. "It is possible to grow plants on sterile sand with purely mineral fertilisers but generally it does not work so well."

Figure 6.1 shows the basic idea of the soil rumen[29]. Using the addition of ammonium nitrate as an example, it shows how charged mineral molecules, called "ions", can be held in the soil and "fed" to the plant and, therefore, the crop. These ions are soluble and move when in solution. In a sand, therefore, they are easily leached away. In a clay some of them, those with a positive electrical charge, are held by the colloidal properties (they act like a sponge, holding ions on their surfaces) of the very small clay particles. The "cations", such as Calcium, Magnesium, and ammonium are held on the soil colloids which carry negative charges. Fertiliser anions, such as nitrate, sulphate, chloride, however, go into solution and are easily leached away. We have historically limited our thinking to "colloids" being clay particles but it is clear that organic matter as humus is also colloidal. And it is possible to improve humus by a number of methods, some of which are hundreds of times more effective than others. With the right method and the right circumstances, the colloidal properties of a soil can be altered as in Diagram 2.2. In other words, it really is possible to change and significantly improve the fundamental fertility of a soil. By increasing this colloidal "ion bank", it is possible to increase yields and reduce costs. Increasing the bank is partly putting extra cash in (mineral nutrients – NPK and trace elements) and partly having a bigger bank to hold more cash.

Understanding how to manage this bank now becomes easier. When we measure soil pH, we generally tend to just think of aiming at a neutral soil, avoiding acidity – so lime is added. However, as Kinsey says, "You can have too much of a good thing. If too much Calcium is added, then the Ca+ ions push other ions off the negatively charged colloid sites and "lock" them up." In effect, they are still there but cannot enter the supply route to the plant root. Crop and soil consultant Peter Wright says: "By just using pH to balance the soil with Calcium, we miss the bigger picture in that other cations (Magnesium, Sodium, trace metal elements), as well as Calcium, influence pH. It is possible to have a pH of 7.5 with the soil still requiring Calcium. Conversely, a soil can have a pH of 5.8 and have adequate Calcium but need Magnesium and Potash to restore the pH to nearer 6.5 which is where soils function best. This is determined by examining the "base saturation" of the soil. Maybe 70% of arable soils in the East of the UK may have too much lime."

Most soils need 60 to 74% of Calcium, 8 to 14% Magnesium, 1 to 3% Sodium and 4 to 12% Hydrogen. Exactly how much Calcium is needed depends on the soil and many other factors, some of which we understand and some of which we don't[20].

Direct Incorporation
In terms of chemical principle, there is little or nothing that can be done in a compost process that cannot be done in the soil. However, the composting process can

reduce or assist in the management of some important aspects of risk.

Firstly, the composting process is at a higher temperature than the activity in soils. This means that chemical, physical and biological risks can be held within the process area, monitored and often speeded up. That is a significant management advantage in managing those risks. However, in reality, concentration of an activity also brings its own risks and the most obvious and common one in this case is the odour so often associated with badly managed composting sites.

Secondly, composting allows a time buffer. Time in industrial terms probably costs money; there will be concrete, machines, energy and manpower involved. However, the first advantage is in holding the material for a suitable crop window. A further advantage is the time factor, this gives the opportunity for the management of processes in a relatively controlled environment and that, in turn, will allow experience to predict the time needed to achieve desired standards and that is helpful in designing Standards which are publicly acceptable.

Spreading

There may be some operational risks when spreading materials which have not been composted. Mineral salts or those which may release ions can be tested for conductivity. Examples of these might be Sodium chloride (which is commonly used as fertiliser for sugar beet) or acidic industrial wastes such as Sodium sulphate (which, again, has fertiliser value). There are also more complex industrial wastes which may contain ionised materials. An example here could be the waste from yeast manufacture where the yeast has been grown on molasses; the residue contains significant amounts of dissolved nitrate. When tested, these will show high conductivity and direct application carries a risk of scorching or burning plants.

There is a more insidious risk in high conductivity materials. They have lower surface tensions and pass through soils more easily and, therefore, increase possible risk of leachate into field drains and groundwater.

Where these risks are evaluated as being significant, the alternative of composting becomes more attractive.

Odour
There is one really important potential advantage of direct application: odour control. Where liquid or solid materials have a significant odour risk, direct application to land, either by injection or by spreading to the surface and incorporating by cultivation shortly afterwards, may make direct application significantly attractive.

How to Manage Liquid Wastes
It is important to remember that the compost process needs the micro-organisms, the food, Oxygen and *moisture*. Green waste on its own, over a year's supply of materials including relatively woody inputs, will certainly go black and friable in most circumstances, but the process is more likely than not to be incomplete with much of the potentially soluble nutrients still soluble. It is only by thorough composting, which needs moisture to push it all the way through its process, that solubles will turn into humus and reduce the leachable nutrients to near zero That is the way to protect groundwater after spreading.

One of the reasons for adopting Deep Clamp composting is that it is comparatively safe to add liquids to the top of a 3m deep heap with very low risk of leachate from the added liquid running through and out of the base of the heap. Regulators with little practical experience of composting will generally argue that the Deep Clamp still needs to be placed on concrete in order to eliminate the point risk. Concrete has significant energy cost and it really is not necessary when a competent operator is given a reasonable Code of Practice for additions of liquids to Deep Clamp operations. Even if those conditions of competence are not where they could and should be, the risks on small scale operations are still very low and are generally out-weighed by the advantages.

If, for whatever reason, the addition of liquids to compost is not an option, then many materials can, safely and at low cost, be spread direct to land.

Direct to Land

To get an even spread and avoid odour, the likely best machine will be a field tanker equipped with dribble bar. (The dribble bar is a wide boom with trailing, large-bore flexible pipes trailing onto or just above the ground.) The bad news about this method is the labour cost and the weight on the ground which, in wet conditions on many soils, will rule it out. An alternative is to use a reel-type irrigator and nurse tank. Again, it is possible to equip the light-weight field bogey with rain gun (which will maximise output and odour control), a boom with sprinklers, or again, a dribble bar (which gives the low odour advantage).

Digestion of Wastes in Mesophylic Processes

Historically, sewage products have caused very significant public nuisance when spread to land. Raw, untreated sewage can no longer be legally spread to land in the UK or most of Europe. Whenever and wherever it is, it stinks. The fist line of processing which many sewage companies use is mesophilic digestion. This allows natural (at UK ambient temperatures) micro-organism process to a temperature of up to around 38°C. There is a very significant reduction in odour and, with sewage and many other organic wastes, there is a usable production of Methane. Indeed, this is the basis of anaerobic digestion.

Generally, anaerobic digestion (AD for short) works well with liquids up to about 10% dry matter. As dry matter increases, the process becomes increasingly difficult to manage and costs rise (from grinding up solids in order to allow the process to proceed). Historically, AD for high dry matter operations has really not worked. Some of the new technology, involving "AAAD" which involves *aerobic* (to get the temperature up), *anaerobic* (to take 60% of the potential Methane off) and then *aerobic* (to get the odour out) does appear to be promising.

Thermophylic Processes

These processes, by definition, involve processes going up to higher temperatures, usually around 60°C to get pasteurisation. In effect, this is what AAAD is. The temperature rise may be produced entirely by natural process or by using added heat to "kick start" the process; this may well be derived by burning some of the Methane produced in the process.

The aim of these processes is two-fold: to reduce pathogens and reduce odour. Generally, the higher the temperature (up to 60°C or so), the more likely it is that reasonable pathogen kill and odour control will be achieved. However, best pathogen control will be achieved at between 55 and 60°C. Also, as temperature rises, so does Oxygen demand rate, and lack of supply will result in a move towards anaerobic conditions with a consequential risk of offensive odour.

How to Manage Biosolids
Sludge and Cake; Nutrient Values; The Sludge Matrix - its value for biosolids and its value as guide to handling MBT output; Environmental Advantages and Safety

Sludge and Cake
"Sewage Sludge" is a phrase with many meanings. In practice, very little original-state (or "raw") sewage gets direct to land in Europe, and many other developed-economy countries, too. When sewage material undergoes a significant level of treatment, it is better and more correctly named "biosolids".

Biosolids sludge is a pumpable liquid with dry matter content usually around 2 or 3 or maybe even over 10% (highly variable in practice). As the dry matter content rises towards 20%, the material becomes increasingly stackable. From around 18% it may reasonably be called "cake".

Nutrient Values
Generally speaking, biosolids are useful in terms of Nitrogen and phosphates but short of potash. Any particular sample can be analysed to give a specific guide. Generally and within the regulations, where biosolids can be applied, they can be applied at a rate which will supply all the Nitrogen and phosphate needs of most crops. Because of the high level biological activity encouraged by the material, it may be that that

activity will facilitate the release of otherwise immobile, or "unavailable" Potash in the soil and that might not be needed either.

The Sludge Matrix - its value for biosolids and its value as guide to handling MBT output.
The "Safe Sludge Matrix[30]" was originally a voluntary agreement between the sewage industry in the UK and the regulators. It was later enshrined into regulatory necessity. The basic principle of the matrix is to try to evaluate actual or perceived risk of disease transfer to humans from a product which, traced back far enough, came from human excrement. The truth is that the risk after secondary and tertiary level treatment is very low but, whatever that actual residual risk might be, there might, in the minds of the uninformed, be a perceived risk. Carrying the public with a project is important in any democracy.

The risks are discounted in an obvious and logical way. For example, raw, untreated sewage is not allowed. Biosolids which have been subject to secondary treatment (such as mesophilic digestion which takes the material up to 38°C) can only be spread to food crops which might be eaten raw (such as salads) up to 30 months before harvest, applied to vegetables which will be cooked up to 12 months before harvest, and to industrial crops (such as oil seed rape for biofuel production) anytime. Further, if the biosolids have been subject to tertiary level treatment (such as thermophilic digestion which takes the material up to 50°C), well, that can go anywhere, anytime.

The system works because it is seen to handle and provide for risk in a clear and workable way.

There is no reason why this approach cannot be used for the output from MBT plants (Mechanical and Biological Treatment). For many years, up to early 2009, Defra prevaricated and avoided allowing the output of an MBT plant to go to farm land, unless the input to the MBT plant was source -separated (which is the whole idea of an MBT that it can take anything and everything, so avoiding source-separation which is expensive, untrustworthy and the public do not like it). It can go to cap out a landfill (so, presumably, there is no risk to the environment). Now, Defra has the compost Standard of PAS100. It is either good enough or it is not. So either adjust it or use it for MBT output. This lack of vision and decisive action to help recycling forward is, at best, sad.

Environmental Advantages and Safety

All organic matter has an ability to hold water. Peat will hold up to 16 times its own weight. Most composts somewhat less. Biosolids materials are able to push composts and soils well towards the moisture-holding capacity of peat. That has advantages in crop production in reducing drought stress and irrigation need. There are also potentially big advantages in run-off and flood control. These "top soil reservoirs" can be built and maintained by adding organic matter in large quantities. Large quantities can be nitrate-safe.

One such source of "top soil reservoir" building organic matter is already within the hands of the water companies; 1 million dry tonnes of output from the STW's (Sewage Treatment Works) in the UK. This will probably give a reservoir effect in an agricultural soil of in excess of 5 million tonnes of water. That original dry tonnage starts off as 25 million tonnes of brown water. If it were thermophillically digested, it would produce methane gas which is a useful biofuel. The residue liquor could be used on arable farms nearby where restricted on irrigation water, it could be piped to farmland close by the point of production. Defibred sludge could be spread direct or stored for a suitable crop window. There are clear advantages to the logistics interests of all parties.

A press report (by the author of this book) in the late 1990's quoted the following:

Charles Booth farms 174ha of arable land near Knottingley, Yorkshire. He has used 4% sludge for 20 years. He "would not like to farm without it". He readily claims lower fertiliser costs, less crop disease and higher yields because of a planned use of biosolids. It is not that he "thinks" that; he has direct, academically credible evidence because Ryhill Farm Services, a crop protection chemical distributor, ran an agronomy centre on the farm with properly run trials. "As an example, one 20ha field had 3920 cu m of sludge applied last August, just before ploughing. That added the equivalent of over half an inch of rain. The land was easier to work and the crop of winter wheat established quickly with less risk. Because of a healthy, vigorous crop, I expect at least 2 tonnes more yield per ha than my neighbours." Independently assessed figures show that this farm has 70% more yield than its neighbours and 80% less fertiliser cost.

The fact is that, partly unknowingly, the water recycling industry is already involved in managing top soil reservoirs. It is also involved in very significant environmental safety advantages compared with other methods of acquiring and applying fertiliser. Biosolids products release plant nutrients slowly, when the crop can take them up, so leaching and pollution of underground water is less than with mineral fertilisers. There is growing evidence of both anti-biotic and pro-biotic effects, so producing healthier crops with less disease and less use of spray chemicals. There is good evidence of better, more consistent yields for the farmer partly because of the soil reservoir effect reducing drought stress on the crop.

David Setler, a researcher in Sri Lanka, said, "Everywhere in the world, there is a need for more reservoirs and dams." Research in the USA gives good, practical data on do-it-yourself (on the farm) reservoir building using old motor tyres as fill for dams and retaining banks.

There is a compelling, common sense argument to build capacity where it is needed. Neither on-farm conventional nor soil reservoirs have the potential to solve all the demand problems but there is scope to make a real and dramatic contribution by reduction of run-off and evaporation. Much of the technology, experience and culture is already inside the water recycling industry. The problem is recognising that fact and using it to advantage. For the first time in fifty years, farming is really ready to help – very actively, if managed right.

Phages and Anti-biotic Qualities of Biosolids
In the 1990's there was a desire, in what we previously called the USSR, to find the next generation of anti-biotics. The logic was glaringly simple. All species have something preying on them. There is the old adage of "every flea has a small flea on its back to bite it….." So, where are there lots of bad guys, because if we can find them, there must logically be good guys preying on them. The obvious place to look, for the researchers at

the time, was in sewage. They had plenty of that and it was already there with a "negative cost". Sure enough, when they looked, they found human pathogens. Sure enough, when they looked harder, they found the good guys attacking the pathogens. The good guys, anti-biotics, were identified as phages. Phages are just sub-optical for the naked eye but can be easily seen with an optical microscope; they look like multi-legged spiders. This, then, is one of the reasons why sewage products can be beneficial in reducing crop disease.

How to Find Suitable Wastes
Sources of Waste; Municipal, Industrial, Business and Domestic Sources; Generic Values of Materials and Values to Land and Crops; The Smart Truck; Specific Materials and Agricultural Value; Values as Feeding Stuffs to Farm Animals; Moisture in Compost Processes; Self Contained Farming; Suitable and Safe Wastes; The Value of Fibres in Soils.

Sources of Waste
It is inevitably the case that when recycling is driven by legislation (or anything else other than demand) then collection and supply of the material will, from time to time, be in excess of supply. At a time when recycling accumulates "separated" material faster than it can be sold or used, it is time to look again at what can and should be recycled to land. Currently, British farmers use around £1 billion's worth of mineral fertilisers, almost entirely imported. Controlled Wastes could, technically, replace most or all of that. Could it be done safely and economically? Would recycling to land be at a lower cost and more sustainable than current separation and collection routes? We may have become too obsessed with sophistication and centralised industrial processes when the most sustainable route has been in use for centuries; proximity recycling to local land.

Back in the early 1990's, the Enterprise Initiative of the DTI funded a series of studies looking at recycling urban wastes to land. Some of these studies ran costs into eight figure but there was one which ran a total under £40,000 before the progress generated began to become self-funding - it resulted in the farmer-owner consortium "Land Network" which, between the individual Members, has, at some point, recycled all the materials in Appendix 3; successfully, safely and within the regulations in force at the time.

Part of that original study looked at how much "waste" there might be nationally which could be recycled to land sustainably. The figures were potentially unreliable but, after many discussions, including with what was then the Centre of Waste and Pollution Research at the University of Hull, the study concluded that there was possibly 100 million tonnes per annum. Land Network now concludes that the figure is higher, maybe much higher. The total land area of the UK is just over 24 million ha but less than 20% is arable and just over 50% grassland and productive grass uses much Nitrogen fertiliser. Forestry would be more productive if compost were applied, too. So, say 10 million hectares could be used for compost substitution for mineral fertilisers. At 25 tonnes per hectare of compost, bearing in mind that composting loses maybe a quarter of its weight, that means that the available land could use in the region of, say 30 tpha of feedstock and a total of, possibly, 300 million tonnes. There is enough land.

Back to the definition of what makes a living organism alive and different from non-living; it is producing a waste. Where there are people, animals in captivity, businesses, factories, there will always be wastes. The concept of "Zero Waste" is only possible in systems where there is a closed loop which gives a sustainable cycle and that is very likely to involve going back to land. So, "the waste business" is centred around chimney pots; where there are people living and industrial activity.

Municipal, Industrial, Business and Domestic Sources
Generally, municipal authorities are a significant collection route for "wastes" and, as far as the European Union is concerned, Brussels has dictated progress by municipal authorities to collect and recycle. The main thrust of much of the regulation has been to direct separation at source, i.e. at the domestic producer household level or at local recycling sites. This "source separation" of municipal solid waste (or "MSW" and otherwise know as bin rubbish or garbage) involves very significant expense and produces poor quality separation which limits remanufacture and delivers an unacceptably low level of safety for recycling to land. This problem of cost and less than satisfactory quality of output will result in developing alternatives which involve collection of whole, unseparated domestic waste. Source separation will fade out as these alternatives become available. That advance will, inevitably, be inhibited by regulation which always fetters any entrepreneurial progress.

Sewage has, historically, often been a municipal responsibility and, therefore, this source of waste has been identified, exploited and regulated. It still has a stigma attached to it in the public and politicians' eyes for, perhaps, understandable reasons and, therefore, there are from time to time, moves to make it "disappear" by gasifying (pyrolysis) it or some other way of extracting energy by incinerating the Carbon in its organic matter. This, of course, is an enormous waste of its potential soil value as a very effective and safe organic manure due to its ability to assist in reduced use of crop protection chemicals.

In Europe generally, most businesses and industries are significantly behind municipal authorities in the restrictions and pressures imposed by the regulators. Nevertheless, because of rising landfill costs, all businesses will become increasingly active in managing their wastes better. It makes sense that raw material should be recovered to reuse in profitable production if possible. The key word is "profitable" simply because handling, processing or doing anything involves resources and has financial implications. The source of money involved depends on the general business environment of prices of raw materials, regulation and related taxation. It is also relevant to think medium and long term in that the climate of what is "acceptable waste" is changing.

Recycling industrial wastes to land is a real technical option limited mainly by regulation. But now is the time to reconsider recycling of a whole range of materials. Everyone in the waste business knows that it is possible to recycle green wastes (from gardens) to land. However, there is a limited supply of that and industry is facing dramatic rises in gate fees and restrictions in going to landfill or to high-tech processing. The plain truth is that the most high-tech processing operation yet designed by man is insignificant in its sophistication and safety compared to the complexity, thoroughness and safety of a compost heap and a fertile soil.

The micro-organisms in a compost heap are mainly bacteria, Actinomycetes and fungi. The soil is of a similar composition but usually with an emphasis on fungi. The numbers in a cubic metre of active compost runs into trillions and the diversity is highly variable (depending on type of material present and the stage of breakdown).

Feed the Bugs
The micro-organisms in compost and land are different from us, of course. However, in terms of dietary needs, they are remarkably similar. They are different in that they can tackle much bigger molecules than we can. Just as we might put sugar (a 6-Carbon molecule) on our cereals for breakfast, the micro-organisms can tackle the lignin in wood (with an almost unlimited number of Carbon atoms in the same molecule) and they never stop eating - 24/7.

As an example of this capability, it is interesting to look at one of Land Network's farms (a consortium of farms which recycle wastes to their own land). The example farm takes wood from local municipal and landscape gardening sources. Cellulose is a polymer made up of long chains of maybe 3000 monomers (groups of 5 Carbon atoms in a ring). So there are around 15,000 Carbon atoms in one of these chains. Lignin is made up of cellulose chains lying next to each other and joined by cross-linkages. This means that xylem (the "wood" behind the bark) in a tree may be almost just one molecule - everything is joined together. The number of Carbon atoms joined together is, obviously, enormous and almost so big that it is beyond ordinary number-comprehension. This farm also takes in PVA - Polyvinyl alcohol as a hot/warm liquid (it solidifies on cooling). As a liquid, it is comparatively easy to disperse it through a large mass of compost (preferably active and, therefore also hot). This is a big Carbon chain alcohol with many thousands of Carbon atoms in each chain. Even so, lignin, the basic molecule of wood, is even bigger with many cross-linked chains. As soil and composting micro-organisms can crack lignin, it follows that they can crack PVA. The BOD (Biological Oxygen Demand – a measure of pollution potential) is enormous, so it is necessary to keep turning the compost (to keep the break-down activity active) but, as a result, the bio-activity rises rapidly and the process is faster. It helps make very good compost. PVA is a plastic; so we can and do compost plastic provided the format allows the micro-organisms to attack it.

So, compost heaps can break down almost any organic molecule - "organic" meaning molecules based on Carbon chains. Given a balance of other necessary food molecules, they will crack 100% of the Carbon chains in the feedstock in time. Furthermore, it does not really matter if there is any residue before spreading to land, provided that the soil is biologically active (meaning that it already has a reasonable amount of organic matter in it).

To say that the Carbon chains will be broken down is not the complete story. The micro-organisms make new Carbon chains in the form of hydrocarbons, carbohydrates and proteins. When the micro-organisms die, these molecules form a dirty black tarry substance (known as DBS) - initially very similar to crude oil. This product is actually what most of us call "humus"; it is the material which gives soil its black colour.

Generic Values of Materials and Values to Land and Crops
Within the context of this book, then, there are a few generalities which give some guidance. If the waste has a lot of Carbon in it, then it can be composted or burned (for energy recovery). However, if the Carbon is present in large chains in the molecules, it is valuable as an

energy source for the micro-organisms. If any useful agricultural crop nutrients are present in the waste, it usually can be composted and recycled to land. These nutrients are the major ones of Nitrogen, Phosphates and Potassium salts, followed closely by Calcium, Sulphur and Magnesium and a whole range of "trace" elements which include most of the "heavy metals" (that regulators generally shy away from).

Evolution has created an ecological balance which can not only tolerate most things, but actually needs most things. For example, as you read this, your blood will contain around 10ppm (parts per million) of Arsenic. However, too little or too much Arsenic would result in illness and probably death. The suitability for recycling to land for agriculture, energy crops, forestry, or land reclamation depends on the composition and dispersal rate of the material in question. This area of technology of environmental management is known as "Dispersion Technology".

There is one other possible way of judging safety; if digestible by humans, it is, possibly/probably, digestible by the micro-organisms in a compost heap. Humans have difficulty in digesting the cellulose in garden wastes but ruminants do not. There is a remarkable similarity between the micro-organism universe of a compost heap and that of the rumen of cattle, sheep, goats, antelope or any other ruminant.

Logic says that the high cost of source separation, so favoured by the EU and the UK government, will eventually be superseded by more dependable recycling routes[32]. If, for example, a member of the public puts a small NiCad battery in a bin and it ends up part of a 10 tonne load of "green waste", there is no way anyone is,

94

in practice, going to find that. What happens in practice is that the resultant compost is spread out far enough to reduce the risk of pollution from that battery to an acceptable level. That can also be achieved without the cost of source separation and the technology for central separation is there, ready and waiting. However, there is one possibility of source separation which is based on individuals in the general public. It is high-tech, packaged and transferable and quite intriguing.

The Smart Truck
At the time of writing, the technology is just beginning to emerge for allowing the public to bring wastes to an automatic separation point where the original product bar coding is used to give good quality separation. Usually, incentive for the public involves "reward points" which can be "spent" in, for example, a local supermarket.

This technology offers the potential to separate out the materials of immediate cash value, leaving the residue for composting and reclamation. It may be, under some circumstances, that there will be no residue for landfill or incineration. (This option will be referred to again under the section on reclaiming arid and desert land in Chapter 9.)

Figure 7.7

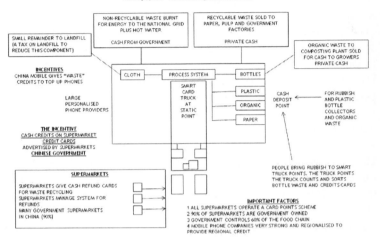

Specific Materials and Agricultural Value

Generally speaking, "Green Waste" from domestic gardens is thought to make ideal compost, but this is not necessarily so. Garden waste from high density housing built maybe within the last 5 years and the waste collected in, say May and June, will be high in Nitrogen and likely to be of high value in composting and to the land onto which it will be eventually spread. However, garden waste collected in, say, October and November, from detached houses built 15 to 20 years ago, is likely to contain large volumes of Cupressus and, if much of this is trunk timber, a lot of energy will be used to shred it and there will not be much Nitrogen in the resultant compost. There is also likely to be a problem in the slow breakdown of the resin in the trunk timber from conifers. If this material is not composted with both another Nitrogen source and significant moisture added, then it is quite likely that crops "fertilised" with the resultant compost will be negatively affected.

Values as Feeding Stuffs to Farm Animals

As mentioned elsewhere, if there is Carbon in a big molecule, then compost heaps and ruminants may well be able to use that material. Generally, if we could eat it, so could a farm animal. Beyond that, ruminants can easily digest cellulose and can, therefore, eat large quantities of fruit and vegetable wastes for food processing factories. Both cattle and pigs can eat dairy product wastes. The question of how far this can be pushed with respect to wastes is not an easy one. For example, in theory, it is quite feasible to feed bioglycerol to ruminants but there are two questions to be answered. Firstly, bioglycerol comes from biodiesel manufacture and if the feedstock for that were used cooking oil, then the bioglycerol is a "Controlled Waste" in the UK and could not be taken to a farm for feeding. If the feedstock were virgin crop oil, then the bioglycerol is not a "Controlled Waste" and could, legally speaking, be fed. This raises the question of whether there is significant residue of Methanol left over from the process (as the biodiesel manufacturing process is methyl esterification based on Methanol with a catalyst). Methanol probably would be metabolised in the rumen but there is little known on how far that can be pushed. Alternatively, the bioglycerol can be composted quite easily.

Moisture in Compost Processes

All composting operations are driven by micro-organisms. And all life as we know it is dependent of moisture. Without moisture in a compost heap, it will result in preservation rather than processing. Composting garden waste without enough moisture will produce a "woody hay". It may be dark coloured and friable but the process of pathogen kill, of immobilizing the soluble nutrients, of killing weed seeds, will all be less than complete.

Having recognised that, there is a practical difficulty in deciding what level of moisture to aim for to achieve the best process. Firstly, materials delivered to a composting site may have too little or too much but they have what they have; adding moisture may be easy enough in principle if it is available (in hot climates, it may not be), and removing moisture could be difficult (such as being so wet that the compost activity has difficulty in starting and generating the temperature that will drive off the excess moisture).

There are two important functions which will dictate where the optimum is; aeration and adequate moisture for process. Firstly, if the input material for composting is well structured, such as with coarsely shredded and woody garden waste, then there will be enough air for the process to start and to build up heat. Furthermore, the windrow or heap will be able to breathe, much as sub-marine snorkels, letting heat out and new air in[9]. From a farming point of view, a range of particles up to 100mm in diameter will give a significant amount of flexibility as to how often the heap needs turning. If there are large particles, then the heap will, in a wide range of circumstances, breathe on its own, without human intervention. As particle size decreases, then that flexibility is lost but the process will progress at a faster rate. Also, where there is a significant proportion of large particles, the material could be very wet and the process still progress satisfactorily. If, however, the material is mainly something like lawn mowings or white onion flesh, then it will slump very easily and run out of air very quickly and go anaerobic and smell offensively.

So, where possible, a mixture of materials will give optimum process control. Commonly, this will have moisture contents in the 40 to 60% range. If the material has good air spaces connected to external air, that moisture content could be as high as 70%, but, especially when poorly aerated, above this temperature, there will be an increase in odour risk. As the moisture drops below 40% there is likely to be process slow down.

There is one area where moisture is in significantly greater demand and that is with some industrially-prepared materials such as MDF (Medium Density Fibreboard). Fibre boards can absorb several times their own weight of water before expansion takes place which disintegrates the board and allows the composting process to proceed[9].

This question of moisture in the compost process will be referred to again below with some test figures.

A tabulated summary of materials, including liquids, which LN has handled is shown in Appendix 3.

Currently, British farmers use millions of pounds worth of mineral fertilisers, with the basic raw material almost entirely imported. All sorts of "wastes" could, technically, replace most or all of that. Could it be done safely and economically? Would recycling to land be at a lower cost and more sustainable than current separation and collection routes? We may have become too obsessed with sophistication and centralised industrial processes when the most sustainable route has been in use for centuries[9]. The old boy in the country cottage garden; he often lived to be over a hundred years old and his garden soil was black with organic materials. He never threw anything away (even what we politely call "night soil"). He recycled everything he did not need to the soil in his garden. He got all the immune stimulation he needed, at all the right levels, all the minerals and all the vitamins. Mostly, it worked incredibly well.

Self-Contained Farming

On the face of it, and logically, a farm ought to be able to support itself without additional fertiliser input, provided imports of nutrients balance export of nutrients in exported crops and animals. If the crop residues and animal manures are all composted down during or at the end of the year, then it is all just going round again. All of it, the trace elements, the Magnesium, the Sulphur, Calcium (lime), the potash, the phosphates and even the Nitrogen. This is even more likely to be true because, in the UK (more in tropical, high rainfall parts of the world) every acre of land gets nearly 2kg of Nitrogen fertiliser out of the rain every year. This comes from lightening flashes in thunderstorms – most of which you will never hear. The searing temperatures of a lightning flash are high enough to force the Nitrogen gas in the air (that Nitrogen is normally inert and will not react with anything) to react with the Oxygen in the air, forming nitric and nitrous oxides. These dissolve in the water in rain to nitric and nitrous acids. These are, of course, very dilute as acids but the Nitrogen is in the form of a fertiliser which the plants in your garden can use.

However, farming usually removes significant quantities of these nutrients in exports. Further, and more insidiously, some of the nutrients, particularly Nitrogen, will leach out in the rain. Not only that, the more the soil is cultivated, the more the organic matter will oxidise and go off into the atmosphere or groundwater. This is particularly true of the Nitrogen in the soil. Nitrate fertilisers will leach out in the rain or during irrigation very easily, even on a clay soil, more so on a sand. Typically, nearly half the fertiliser applied, if it is mineral Nitrogen, leaches out. If the Nitrogen is in the form of the proteins in humus, it will not leach out, but if you cultivate the soil, it will tend to oxidise. As that happens, if the soil is moist, the Nitrogen will turn to nitrates and be easily lost into the groundwater. In very hot, dry weather, it may go off as ammonia into the atmosphere.

It is over-simplistic to conclude (but there is a point here) from this that the worst thing a farm can do from a nutrient maintenance point of view, is cultivate. The alternative is to mulch and let the worms do the work.

Wastes Which are Suitable, and Probably Safe

Repairing the loss is relatively simple; put nutrients and organic matter back into the soil by importing waste, preferably organic matter, from outside the farm.

Green Wastes: including lawn mowings, bedding plant residues, hedge trimmings, fresh and cooked vegetable waste; Good source of organic matter that will help to make humus. Green leaves especially valuable because they contain Nitrogen. Lawn mowings in the spring are the "dynamite" of composting with much Nitrogen and moisture. On their own, they will use up the Oxygen quickly and "slump" down into a wet, soggy mass and are likely to go anaerobic and smell. Mix mowings with other material and move it more often. In the autumn, remember that woody materials have much less Nitrogen in them.

Kitchen Wastes: Stop! There is a legal problem. If it contains meat residues, or might do so, then that material is covered by the Animal By-Products

Regulations (ABPR) and that means composting or other treatment in isolation – in a building of some sort, sealed off from birds and vermin.

However, if it can be handled, inside or out, then the micro-organisms in the soil and the compost heap are very much like ourselves in terms of nutrient needs, but they can, given time, digest almost anything. Below is a list of common kitchen wastes and their values.

Uneaten food after a meal; Vegetable waste as above, meat good (contains much Nitrogen) but chop it up as it is slow to disintegrate.

Paper and cardboard; Not the best but by no means useless. Paper products are mainly Carbon which means that the micro-organisms can use it for energy and as the basic building block to make humus. However, to do that, it will be necessary to also feed some Nitrogen, in particular, and the full range of other plant foods including trace elements. These can come from the materials above in this list but don't over-do the paper. How much is "over-do"? Well, it is partly experience but the technology will give you the framework. (See "Carbon:Nitrogen Ratio below.)

Ash from fires; Great if it is wood ash; that will give much potash and a good range of minerals (or "trace elements"). If it is coal ash, that will yield up the minerals but very little potash. All black coloured ashes have another effect (see next paragraph).

High Carbon Ashes, Charcoal; If the ash is very dark, really black, then it may have another value with a long term effect.

Studies of black soils (called *Terra preta*) in South America[35] [36] have put another angle on Carbon and its value as amorphous or activated Carbon, usually (it is thought) derived from charcoal. This charcoal is sometimes now trendily called "Biochar". *Terra preta* soils show a very high degree of sustainable fertility which, in areas of high temperature and violent fluctuations in soil moisture, might at first appear surprising. Initial studies on Land Network farms indicate that the Carbon in printing inks appears to be of similar value. The Carbon is capable of holding onto nutrients at a level similar to humus. However, humus is easily and rapidly oxidized under conditions of high temperature and moisture but the pure Carbon is not. It may be that there is potential for the wider use of Carbon in soils in ways that we do not yet fully understand.

So, dark ashes are more likely to be beneficial than not.

Animal manures; Great stuff with plenty of food for the micro-organisms and a fresh injection of its own micro-organisms (which may speed up the whole process). Horse manure is a useful source of potash. Poultry droppings are very rich in Nitrogen. Cattle manure is good in phosphate and Nitrogen. Land Network has even handled Elephant dung. Yes, and the dog and cat can contribute, too. With all of these, because they potentially carry human pathogens, getting the temperature up is important, so do use a thermometer.

Values of manures vary from species to species, with what they are fed on and whether there is bedding with the dung and urine. Some guides are as follows and these are without straw or other bedding;

Manure

	Likely % Moisture	Percentage in Dry Matter N	P	K
Cattle	75 to 85	2	1	0.5
Pig	75 to 85	4	2	1
Chicken	60	6	6	3
Sheep	60 to 70	4	2	1

Animal bedding; The cat litter may be volcanic ash which is inert and will add to the structure of clays. Sawdust is mainly Carbon (useful for humus production provided there is at least some droppings and urine). On this latter material, in old Victorian gardens, the old gardeners would encourage the apprentice gardeners to "pee" on the compost heap. Human urine (and of other animals, too) contains uric acid, which contains Nitrogen. Urine is (because of the acid) mildly anti-septic; so it is unlikely to carry disease if it is fresh urine (best not to leave it hanging about - get it into the compost heap).

Carpet shredded to a "fluff"; Great stuff, especially if it is from wool carpets. Even if it is from synthetic fibres, these will be very valuable as fibres stabilise the physical structure of soils and allow them to breathe and manage water better.

And Some More Difficult Materials:

Cupressus prunings; It is possible to compost these but they will process faster if they are shredded. All the coniferous trees, however, contain resins and these are very difficult for the micro-organisms to break down. Turn more often.

Wood and MDF; The wood is mainly Carbon and will need chopping or shredding into fairly small pieces.

MDF is very interesting. It is, of course, predominantly wood. There is, generally, a lot of misunderstanding surrounding MDF, chipboard and similar process wood sheet materials. Such materials are present in significant quantities in many of our homes, offices and the confined spaces in which we live. The most common binder or glue is ureaformaldehyde. Urea is used in large quantities, world-wide, as a Nitrogen agricultural fertiliser. The second part of the molecule, formaldehyde, _if it were separate and on its own,_ is known to be toxic and potentially carcinogenic. The combined molecule is **not** toxic. (Sodium is a silvery metal which dissolves in water and bursts into flame as it does so; if it were eaten, it would burn a hole though the body until it came out the other side. Chlorine gas is quite poisonous and a few whiffs will kill.)

Ureaformaldehyde is used both as an agricultural fertiliser and has been used for many years as a slow release Nitrogen source in many general potting composts sold in garden centres[33]. The key phrase is "slow release" which has significance both in commercially in growing and environmentally in reducing pollution.

So, if people tell you this material is dangerous, ask for their evidence and suggest that, if they were to be right, we all have much more risk from the furniture in our kitchens and offices. WHO (the World Health Organisation of the United Nations) has had a look at this and there is a clean bill of health.) Everything has risks but, as far as we know at present, any risk from MDF is very, very low. The risk as a fertiliser is much lower that the risk in a typical kitchen or office.

So, how can you use MDF in composting? Well, it is necessary to let the bugs get at it. One way is to shred it (which takes quite a lot of energy) or break it up a bit (say 15cm bits) so as to leave a broken edge. For tree relatively large pieces, say 150mm square, can be buried under permanent planting for, forest trees or land reclamation in preparation for a permanent planting. That will give a slow release of Nitrogen. Alternatively, shredded material, or just broken up into 150mm squares, can be mixed with some farmyard manure or active compost made from this or other materials (these will add some micro-organisms and "seed" the process), then make sure it is quite damp and kept that way. A 15mm thick sheet will keep on absorbing water until it is 50 to 100mm thick. Then it will break up and can be mixed in with the compost or spread direct to the ground. It is a really good source of Nitrogen – so, until you are used to it, not too much in one place and allow moisture and time. Broken MDF boards, mixed 50/50 (ish) with cattle slurry will break down reasonably quickly – but still maybe 2 or 3 times longer than if it were shredded. (If time is not a problem, this route is, of course, less expensive on energy.)

Plasterboard; Plasterboard is a sandwich of Gypsum, which is Calcium sulphate, between two sheets of paper. Plants need both Calcium and Sulphur. The Calcium is especially useful in helping flocculate clays to give a better crumb structure, make them easier to work and plants will establish faster and grow better. However, the Calcium sulphate is fairly alkaline and so use sparingly if adding to a compost heap. Avoid plasterboard with an Aluminium foil backing; that will not break down. The same remarks apply to waste plasters – most of which will be Calcium sulphate based. Better technically (although UK regulations prohibit this) to put all of these direct to the soil where broken up plasterboard will break down as it gets wet but it will look untidy in the meantime.

Old mortar; Old mortar from an old wall will have been made from lime (probably "quick" lime which is Calcium oxide) and sand. Crops need the Calcium; it will help the compost, the soil and the plants. However, the mortar does need breaking up, preferably sieving, and remember how much it is diluted along the way onto the soil will affect the following plant growth. Calcicoles like it; calcifuges don't! Mortars are also alkaline and, therefore, go sparingly if adding to the compost heap.

Coffee grounds; Yes, the compost process and the soil will deal with this, too. Spread it out a bit or it will keep the micro-organisms up all night.

Teabags; Have you ever thought how tough a teabag is? You put it in boiling water and push it round violently with a spoon and even the couple of millimetres round the edge does not give in. How do they "glue" those edges to stand that? Why does the paper not disintegrate and let the tea leaves out? The answer is that this is high-tech paper. There really is quite a lot of advanced technology required to design and manufacture a teabag. The base material is paper but the strength comes from the addition of a polymer plastic. Most gardeners will already know that teabags are slow to break down; this is why. Is it bad or toxic? Not a bit of it. The plastic is composed of just Carbon and Hydrogen. It will take time to break down but it will break down and do no harm. The fact that it is slow to disintegrate is a benefit because it adds fibres to the soil.

New and used carpet; Regulators have been very slow, wrongly so, to allow farming to recycle carpet to land. However, gardeners have used carpet over many years to wrap round a compost heap to allow the process to breathe while letting rain in and reducing heat loss. These are all very important assets of carpet which is used as a carpet; just as a piece of material, still recognisable as carpet and used in pieces of a metre or two across. Gardeners have used the "fluff" from a vacuum cleaner for years by adding it to the compost heap and spreading it amongst the other materials.

All of that is a really useful application for carpet and it will eventually break down. But suppose we accelerated that breakdown by tearing it up. What is its potential value?

The majority of carpets sold in the last 20 years in the UK have been 80:20 wool:nylon. The backing has moved from jute, or hessian, to polypropylene. The immediate response is that it is the wool which is valuable. Wool is certainly of value because it is protein and that contains Nitrogen. However, there are other things of value. Look, for a moment, at the following analysis.

Table 7.8

Typical Breakdown by % Weight of Average 80:20 Wool-Rich Carpets

80:20 with pp backing			% of total
facefibre	35%		
wool		80%	28%
nylon		20%	7%
adhesive/filler	45%		
SBR		20%	9%
chalk		80%	36%
backings	20%		
pp		100%	20%

80:20 with jute backing			% of total
facefibre	35%		
wool		80%	28%
nylon		20%	7%
adhesive/filler	45%		
SBR		20%	9%
chalk		80%	36%
primary backing	5%		
pp		100%	5%
Secondary backing	15%		
jute		100%	15%

(SBR is synthetic latex, and the PP (polypropylene) fibres are not UV stabilised.)

Now, here is an interesting thing; the biggest single component is chalk! All plants use Calcium, the main value of chalk in soil. By the way, the adhesive will break down releasing energy to the micro-organisms. The nylon and propylene have real value and will be discussed in some detail below. So, carpet has a value in providing Nitrogen, Calcium and fibres.

Woollen fabrics; Wool is pure Keratin – a range of proteins. Proteins contain Nitrogen in big molecules. Big molecules do not leach out, so this is slow release Nitrogen. So, old woollen sweaters? "Plant" one under a shrub or perennial. Alternatively put it in a compost heap; it will be easier to turn the compost if you cut the woollen fabric up a bit before adding it to the heap. It will be slow to break down but an active, warm heap will do it. It really does not matter if it does not break down completely.

The Value of Fibres in Soils
The last few paragraphs above have revolved around the value of fibres in soils. So, what is that value?

It is important to remember that all of the basic physical structure of soil is composed of mineral matter which is generally regarded as not biodegradable. Most soils, as any textbook, and the Environment Agency's publication "Understanding Rural Land Use" indicates, will also point out that the structure of good, stable, really fertile soils also depend on plant roots and "organic" matter to create a matrix which gives further characteristics to the mineral matter in soils in terms of stability, gas exchange, moisture management, cohesion and erosion stability.

It can also be observed that many hydroponic crop production systems use vermiculite, polyurethane and many other synthetics as a growing medium to provide a matrix for root growth and a structure with a large

110

surface area which the mycorrhiza can latch onto. These materials will normally be described as "not biodegradable". This is not strictly true in that all of these fibres be they "natural" (such as lignin from plant roots) or "synthetic" (such as polypropylene from carpet backing) are all biodegradable but the process will, fortunately, take much time, maybe years.

Now, as any one who has any knowledge of growing plants knows, peat is used as a growing medium. "Organic matter" is regarded as providing stability and a range of growing qualities to soil. These materials contain cellulose, hemicellulose, holocellulose and lignin. All of these are big molecules, larger than the synthetics in the carpet mix. It is the fibres in peat which are the secret of what it imparts to soils it is mixed with or to a growing or potting compost.

Discussion on the Analysis of Wool
Wool composition is of "Keratin" proteins which, of course, contain Nitrogen and, in the case of keratins, useful amounts of Sulphur. In proteins of the Keratin group, several of the amino acids contain Sulphur and, of course, so do the amino acids which form some microbial proteins, plant proteins and animal/human proteins. These nutrients are clearly large molecules and therefore the nutrients are slow release and on crop demand.

Most people don't know this but, in most UK carpet composition, the most common material in the carpets is, as shown above, chalk which is used as a filler and stabiliser for the latex which glues the tuft to the Hessian or polypropylene backing. Again, the chalk contains Calcium which is a secondary, but very

111

important, nutrient; second only to the major nutrients of NPK.

Jute used to be (and still is in what we will get for several years until currently new carpets get replaced) the most common backing and this is a fibre of plant origin, supplying Carbon and energy to soil micro-organisms. Cellulose is, of course, mainly Carbon molecules in long chains of 5-Carbon rings. Woody stems, as in jute, will contain larger molecules of hemicelluloses with these long chains joined by cross linkages. There may also be some lignin. Lignin contains many hundreds of 5-Carbon ring, cross-linked chains. Lignin is closely associated in woody materials with cellulose and hemicellulose. The point here is that these "natural" fibres contain many hundreds of joined-up Carbon atoms. Most people accept that these fibres will take some time to break down in natural environments of soils or compost heaps, but these same people rarely question how long degradation takes.

The "synthetics" are large carbon molecules - actually with smaller chains than in many natural fibres (see above). In the case of carpets, they are not "stabilised" against degradation by Ultraviolet light; so they are a little easier for the micro-organisms to digest. There is no question about breakdown and degradation; these do occur. The questions are how long this takes and, with this material, in this application, at these dilution rates in the garden, is a difficult thing to be precise about. Indeed, straw ploughed in may last several years if the soil is not "breathing" to that depth. Whether slow breakdown is a good thing depends on circumstances but there are few or no known, likely or measurable deleterious effects.

The tufts contain, usually, 10% of fine nylon filament fibres. Nylon does have a very small amount of Nitrogen in its molecule. We do not yet have detailed knowledge based on long term experience but the technology indicates that these fine filaments, when composted, and dispersed, will undergo significant degradation in maybe weeks rather than months in the soil. (Years are very unlikely.)

Polypropylene (Unstabilised). Table 7.8 on page 110 shows two analyses. The first is of jute-backed carpet – the majority of post-consumer carpet is like this. New carpets are like the second – for carpets which are "post consumer" (taken out of the house when replaced by new carpet) we expect a progression towards this material over some years. Ten or fifteen years ago, there was little polypropylene used in new carpet backing – it was all jute (or hessian). Now, the situation has reversed. As these new carpets age, it might be expected that the percentage of polypropylene in "waste" carpet will go up. However, there is technology which is developing which can separate the synthetic from the natural fibres and there is an incentive in that recovered synthetics can be worth over £400 per tonne at current (2009) prices. So, it is reasonable to think that the percentage of polypropylene may go down, rather than up, as the new carpets come up for replacement. What there is more caution about is how long polypropylene, if and when it is present, will take to break down under these circumstances.

Why all this on carpets and fibres? Well, fibres are an important part of the structure of soils. Cultivation will accelerate oxidation of these large molecules and the soil will breathe less, water will flow through less, there will be less water held in the soil in a "healthy" way and there will consequently be more drought stress. Fibres make soils more productive and plants less subject to diseases. Erosion of soils by air and water is a significant factor in the loss of soils in all arable farming systems, all over the world and the UK is no exception. The UK Environment Agency recommends[37] the build up of organic matter in topsoils so as to stabilise them against erosion, reduce surface run-off and thereby reduce the risk of flash flooding.

Municipal and Industrial Wastes
Just, for a moment, look farther than your boundary fence and look at the national picture. Currently, Defra (in its great wisdom) is encouraging local authorities to "recycle" wastes through large industrial processing plants; Energy from Waste[38] [50]. Whether you think that this is incineration by another name (which it is), and whether you think the energy gained means we save a small amount of burning fossilised fuels, is a discussion which appears to have been lost somewhere. What the route does do (because the installations the UK is going for are very large ones) without any doubt, is centralise waste collection, maximise trucks on the road and destroy organic matter which gardeners, nurserymen and farmers could grow flowers, food and biofuels with, so to save imports of mineral fertilisers and get trucks off the road by proximity principle recycling. There appears to be a misguided obsession with large scale operations because of a blind assumption that there is economy of scale. It is interesting to note that comparatively small towns in Scandinavia may have two or more energy from waste plants. They are flexible and can change relatively easily with market conditions and they get trucks off the road. Heat is comparatively easily piped into local homes. Perhaps most importantly, the local people see them as "their" property.

Farming is closer to recycling on a "Small is Beautiful" or Proximity Principle. There are a lot of farms and if farmers do it, and the farmers, and the foresters, and the national parks, and all land users who currently use mineral fertilisers, yes, the pollution of groundwater can be eliminated and infringement of the standards laid down in the Nitrate Directive avoided.

Municipal and Industrial Recycling
Most businesses and industries are significantly behind municipal authorities in the restrictions and pressures imposed on local authorities by the regulators. Nevertheless, because of rising landfill costs, all businesses will become increasingly active in managing their wastes better. It makes sense that raw material should be recovered to reuse in profitable production if possible. The key word is "profitable" simply because handling anything, processing anything, doing anything, involves resources and the cash to pay for those resources has to come into the equation somewhere. That source of cash depends on the general business environment of prices of raw materials, regulation and related taxation. It is also relevant to think medium and long term in that the climate of what is "acceptable waste" is changing. Whatever anyone thinks about the word "profit", it is as well to remember one principle; if a situation is environmentally not sustainable but is financially sustainable, then it can go on for hundreds of years, but if the reverse is the case with a situation being environmentally sustainable but not financially sustainable, then it dies today. The fact is that true sustainability involves both environmental and financial balance.

Recycling industrial wastes to land is a real technical option limited mainly by regulation. Now is the time to look again at recycling the what, when, where, who and how - the why is a rhetorical question. Everyone in the waste business knows that it is possible to recycle green wastes (from gardens) to land. However, there is a limited supply of that and industry is facing dramatic rises in gate fees and restrictions in going to landfill or to high-tech processing. The plain truth is that the most high-tech processing operation yet designed by man is nothing, nothing to match the complexity, thoroughness and safety of a compost heap and a fertile soil. So, whatever the Local Authority does, whatever technology is developed to do clever things in a big shed on an industrial estate, it is still infinitely preferable to keep trucks off the road and recycle as much as possible to farm land on a proximity basis.

How to Manage the Nutrients in Waste
All materials, not just wastes, contain, at least to some extent, "food" for micro-organisms. Natural ecosystems are insidiously capable of reconstructing almost everything through closed loop processing. Even the Titanic, sunk two miles down, is not rusting in the absence of enough Oxygen, but it is being eaten by bacteria. It may take several hundred years, but they will do it. This demonstrates something of regulatory significance. Very often, those not skilled in the art, with limited technical knowledge and/or experience, will say something won't work or is impossible. Most things work in nature; it will handle anything given time. The question is more about how long it will take.

There is a word of caution in thinking about Carbon and its use in soils. Carbon certainly can be involved in large molecules which are, in the language of an organic chemist, "organic". Carbon may also have a very significant effect on the fertility soils when it is present

116

as pure Carbon, in its amorphous or activated state, as shown in the *terra preta* soils of the South Americas[43][44].

Micro-organisms certainly need food and they need moisture and gas exchange - if we are interested (as we are) in aerobic composting, they need Oxygen from the air and need to be able to breathe out Carbon dioxide, just as other forms of life, including humans. But what do we mean by "food" and how flexible are they? Bearing in mind that farmers in the UK have been ploughing in 6 to 8 million tonnes of straw per annum for over 20 years in the UK [25], does a text-book view of "Carbon:Nitrogen Ratio" really matter?

The "food" that micro-organisms need is basically Carbon (for energy – just like sugar and carbohydrates are to us) plus all the other nutrients that our own bodies live on. The Carbon can be in the form of almost any organic molecule (i.e. any molecule containing Carbon linked to other Carbon atoms, Hydrogen and possibly many other elements). They also need a balance of Potassium, Phosphate, Calcium, Sulphur, Magnesium and the full range of trace elements, just as we do. Micro-organisms do differ slightly to humans though in their ability to use a wide range of Nitrogen sources to build their own body proteins. Unlike humans, they can use non-organic forms including, in the case of the bacteria attached to the roots of legumes, atmospheric Nitrogen. They also have the flexibility to be able to adjust to wide variations in the concentration of organic Carbon molecules.

In discussions about C:N Ratio, a range is often given, commonly 20 to 30 to 1, without reference to whether this refers to the start of the compost process or where the farm wants to end up with in the finished compost.

There are many text books, academic papers and standards relating the proportion of Carbon to Nitrogen in a compost process and product[41].

Basically, micro-organisms need a lot of Carbon (for energy) and not so much Nitrogen (to make cell body proteins). Many "authoritative" guides will give, as a guide, an ideal of 25 of Carbon (by weight) to 1 of Nitrogen. Alternatively, a range of 20 to 30 to 1 is usually a better guide to what actually works. The truth is that it can be much wider than even that and this is because the micro-organisms will use up some Carbon in the process. Furthermore, the more difficult the process, then the longer it will take and more energy, and therefore Carbon, they will use up.

As a guide, the following gives an indication of the C:N ratio of a number of materials which may be "on offer" to a composting operation:

Material	C:N Ratio
The best top soil	10:1
Humus	10:1
Kitchen food waste	15:1
Vegetable wastes (factory)	20:1
Lawn mowings (UK May)	10:1
Lawn mowings (UK August)	25:1
Tree leaves (autumn brown)	35:1
Manure (stable- high straw)	50 to 70:1
Manure (cattle, low straw)	20:1
Cereal straw	70 to 100:1
Sawdust	150 to 800:1
Newspapers	150 to 200:1

As an example of handling high Carbon inputs successfully, in the mid 1980's British farmers faced a ban on burning cereal straw behind the combine harvester. The question was: what would happen when around 6.5 million tonnes of straw was ploughed in every year? Put another way; how could this be best managed[25]? By the mid 1990's, every farmer was doing it without a second thought and yet the straw had very little Nitrogen in it and the C:N ratio was well outside the figures discussed above; straw will, depending on variety and season, have a C:N Ratio into the 70 to 100 to 1 range.

Generally speaking, decomposition by micro-organisms is much the same process as digestion by ourselves in that it requires energy. As we metabolise sugar during exercise, they will metabolise Carbon-based molecules to produce the energy to drive the process and produce Carbon dioxide. They need Nitrogen to build cell proteins but they need much more Carbon than Nitrogen. In processing the Nitrogen, they will metabolise Carbon and the more difficult the digestion, the more Carbon they will use up. As the C:N ratio rises above 30:1, the process gets more difficult and uses up more Carbon and takes longer.

There is one more factor to complicate the issue and the judgement on how well the process will go and whether more Nitrogen will be needed. With ploughing in straw, for example, some of the Carbon is in the form of cellulose and will quickly break down if there is enough Nitrogen in the system. Further, there will be a lot of Carbon present as more complicated molecules with more stable cross linkages in the Carbon chains, such as hemicelluloses and lignins. These will take longer to process and may not require more Nitrogen for the micro-organisms in the short run.

Research up to the point when straw burning in the field was banned in the early 1980's in the UK showed clearly that ploughing in too much Carbon, too soon, would produce what farmers called "Nitrogen Starvation". This meant that the soil micro-organisms would take Nitrogen out of the surrounding soil and use it for processing the excess Carbon, thus robbing the soil of Nitrogen for the current crop which would, then, suffer. However, the soil usually recovered after two or three years and the rules, then, for incorporating straw involved[21] adding some easily available Nitrogen supply after the first year of straw incorporation, a little less in the second year and maybe some more in the third. Then the soil would cope. These rules would be modified with less Nitrogen added if the soil had significant biological activity (because of previous organic matter additions encouraging the populations of mycorrhiza) or increased and lengthened on arable soils depleted of organic matter (and, therefore, the activity of mycorrhiza).

Interestingly, if the supply of Nitrogen is less than the C:N ratio of 20:1, (i.e. the material is Nitrogen rich)then the soil will increasingly ammonify the Nitrogen compounds. If the soil is wet, this may then be washed into the groundwater, in dry, warm conditions, the ammonia may come off as a gas to atmosphere. Thus another reason, in a composting operation, not to let the heap dry out. With adequate moisture, and preferably with the ultimate addition of enough Carbon, the micro-organisms will find a way to convert that Nitrogen into large molecules which become part of the humus and, therefore, not leachable and not likely to be released to atmosphere.

Different C:N Ratios will produce different compost temperature curves and different processing times. In part, this can be managed by altering the turning regime, but only in part.

Figure 7.9

Figure 7.9
EFFECT OF CARBON:NITROGEN RATIO ON COMPOST PROCESS

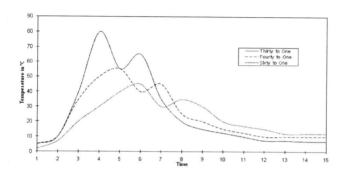

Generally, low Carbon/high Nitrogen materials will produce higher temperatures and process faster, while high Carbon/low Nitrogen materials will be slower and reach lower temperatures.

These rules are, of course, generalisations in a complex environment. For example, if the Carbon is present as cellulose, the rate of metabolism by the micro-organisms is potentially much faster than if it is present as hemicelluloses and, again, that too would be faster than if the Carbon were present as lignin (i.e. wood). Metabolism of high Carbon materials, especially these "tough" woody materials, will use up Carbon. This means that if the feedstock for composting is of very woody material, then the C:N ration could be high, maybe much higher than 60 or 70:1 and there would be little more detrimental effect then than taking longer to process. This may mean that input materials may be above even 70:1 provided they come down nearer the

121

30:1 by the end of composting. Despite this, some heavy land farmers[9][10] are, correctly, happy to see woody particles left in the compost, knowing that they will break down slowly, in that they will help to hold a clay "open" and let it breathe better, thereby allowing faster soil warming in the spring, better gas exchange with less resultant chlorosis, better moisture movement and, overall, better crop growth. Where the remaining Carbon is as lignin, then it will only slowly be metabolized in the soil and is unlikely to significantly affect the supply of Nitrogen to a crop. The best advice to farmers[9][25], is to apply the compost in any way possible and, in the early years of converting to compost farming, watch the chlorosis in the crop and use instinct to top dress with mineral Nitrogen fertiliser if necessary. When well into compost farming, the biological activity in the soils will mean that that precaution is unlikely to be necessary.

In commercial operations where the objective is a low priced material which can legitimately be called "compost", then it may be that the uninformed will sell composts where the C:N Ration is well above 30:1. Such a compost may be friable and dark in colour but that does not mean that it will support plant growth at all well. These composts may well turn out to be less than reasonable in performance and seedlings planted out into it may fail, quite likely, completely. It is reasonable for the purchaser to expect plants to grow and, therefore, that product could reasonably be "unfit for purpose" in law. This, of course, is the reason for Standards and the current UK Standard, PAS100, which is, unfortunately, the only one. It is not particularly suited to farm requirements and it has a very restricted range of permitted materials and, therefore, inhibits recycling to land. However, at least, it is a start.

If the value of the input materials is known, then it is possible to predict the value of the output in terms of value in the soil and to a plant or crop. As a guide, composting micro-organisms will use about 30 parts Carbon for each part of Nitrogen, and such an initial C:N (available quantity) ratio of 30:1 promotes rapid composting. Researchers report optimum values from 20 to 31:1. A majority of investigators[45] believe that for C:N ratios above 30:1 there will be little loss of Nitrogen. University of California studies on materials with an initial C:N ratio varying from 20 to 78 and Nitrogen contents varying from 0.52% to 1.74% indicate that initial C:N ratio of 30 to 35 was optimum. These reported optimum C:N ratios may include some Carbon which was present as hemicelluloses or lignin. Composting time increases with the C:N ratio above 30 to 40. If proportion of large molecule Carbon is small, the C:N ratio can be reduced by bacteria to as low a value as 10:1. Fourteen to 20:1 are common values depending upon the original material from which the humus was formed. These studies also showed that composting a material with a higher C:N ratio would not be harmful to the soil, however, because the remaining Carbon is so slowly available that Nitrogen starvation would not be significant.

Does all of this really matter? Well, the C:N Ratio will certainly affect the rate of composting process and how plants will grow if planted into pure, undiluted compost. However, seeds and plants in agriculture are not planted into pure compost. In this case, it is how the addition of compost will affect the soil as a whole and into which a crop will be sown and grown. If the technical base is assumed to be purely chemical, then it may be that an uncomplicated view of the chemical value of the compost is a sufficient guide. However, the technology indicates that there is a wider biological effect and the Carbon input into an arable soil with run-down organic matter may well benefit from higher inputs of Carbon because the soil mycorrhiza will use up Carbon in their energy systems, just as is the case with the composting organisms.

There is one area of use of large Carbon chain molecules in composting which has been sadly neglected in developed agriculture and with results which have sometimes been visibly catastrophic but, maybe universally, have allowed a progressive and often un-noticed erosion[37]. Fibres of root hairs, lignin, and synthetic Carbon molecules all act as stabilisers of the physical structure of soils. They help to allow Carbon dioxide from soil micro-organism activity out of the soil and Oxygen back in. They allow water into the soil (reducing surface run off and flood and erosion risk). They help hold water (reducing irrigation need). They reduce cultivation need and energy used in cultivations when they are needed. In terms of soil physics, it does not matter if the Carbon molecules are synthetic or "natural", nor does it matter how long it takes to break them down – in fact, it is better if they just stay there.

Inorganic Carbon
Studies of black soils in South America has put another angle on Carbon and its value as amorphous or activated Carbon, usually (it is thought) derived from charcoal[43][44]. This charcoal is sometimes now trendily called "biochar". A study at the University of Edinburgh is looking at the potential to use biochar as a means of locking up Carbon from the atmosphere in a fairly permanent way and so as to make soils more fertile at the same time.

Terra preta soils show a very high degree of sustainable fertility which, in areas of high temperature and violent fluctuations in soil moisture, might at first appear surprising. Initial studies on Land Network farms indicate that the Carbon in printing inks appears to be of similar value. The Carbon is capable of holding onto nutrients at a level similar to humus. However, humus is easily and rapidly oxidised such conditions of high temperature and moisture but the pure Carbon is not[43][44]. It may be that there is potential for the wider use of Carbon in soils in ways that we do not yet fully understand.

How to Manage the Bugs in Compost
The basics are simple: food, moisture and Oxygen. Continuous correct levels of food, moisture and Oxygen, will produce staggering results.

Soil Mycorrhiza, Crop Diseases and The Closed Loop
Glomalin - the binding agent in soil structure; Plant Metabolism; Anti-biotic effects - crop diseases; Pro-biotic effects - nutrient release and crop health; Glomalin - the binding agent in soil structure.
Some years ago there was a breakthrough in identifying soil "glue" and its relationship to soil crumb structure, soil tilth, gas exchange, soil water movement and retention, crop stress and disease. Researchers at the Soil Microbial Systems Laboratory, Beltsville, United States Department of Agriculture (USDA), identified a protein called Glomalin which appears to be the "glue" which holds soil aggregates together. This breakthrough can be coupled to advancing knowledge on mycorrhiza and can open up the route to lower costs in cultivations, lower costs in crop protection and higher crop yields.

Sara Wright[19], a microbiologist researcher at the USDA, was part of the team which identified and named "Glomalin" as the protein which appeared to be the binding agent in the formation of soil aggregates and, therefore, the controller of so many practical functions in which the operator in the field is so interested. The team found that tillage tends to lower Glomalin levels.

Soil Aggregation

During the 1980's, quite a bit of academic research was done on soil aggregation and the formation of crumb structure. It was clear then that cultivation increased the mineralisation of organic matter which was closely related to the formation of aggregates. In turn, it was also clear that such loss of aggregation also resulted in reduced gas exchange, reduced water movement, reduced crop growth and increased power requirement for cultivations. As many arable farmers know to their cost, the pressure to get on and do the job gets the operation onto a descending spiral of declining soil structure, pressure on yields and increased costs.

Research also confirmed that what grandfather knew was indeed true and now could be measured in scientific terms; i.e. that some crops were more capable than others of putting organic matter back into the soil. Hence the historical practice and interest in "green manuring" (growing a green-leaved crop and ploughing it in) and, more recently, ploughing unwanted cereal straw back in. The research led to the "Hierarchical Theory" which proposed that soil stability depended on macro aggregates which were based on micro aggregates held together with organic matter such as pieces of roots, stems, leaves and other plant remains. Hence, the interest in cultivations which resulted in the breakdown of organic matter. In turn, the micro aggregates were held together by small, negatively charged clay particles. That was clear enough but it has become progressively apparent that the binding of micro

126

aggregates into macro aggregates is not only dependent on active plant roots and decaying plant remains, the presence of fungal hyphae is also a fundamentally important part of this activity. The identified fungi are Vescicular Arbuscular Mycorrhiza - VAM fungi for short. There is no doubt that the biological activity of mycorrhiza is of fundamental significance in terms of science, technology and commercial farming.

The function of fibres in the soil is interesting; in a peat there are fibres of hemicellulose and lignin which may be hundreds or even thousands of years old. These clearly have a physical value in gas and water management when mixed into soils. There is some logic and practical evidence that synthetic fibres, such as nylon and polypropylene from carpets, can and do have a similar function and value.

Plant Metabolism
These VAM fungi are not just involved in the physical characteristics of the soil, they affect, are directly involved with, and are likely to be the major route, for the nutrient uptake by the plant. There is a very close relationship between VAM hyphae and plant roots. Normally, all plant roots are covered in VAM as symbiotic assistants. They (VAM or roots) are known to be deeply involved in the uptake of phosphate but there is evidence that they also increase the absorption of Sulphur, Magnesium, Iron, Zinc, Copper, Manganese and probably most other things which the plant needs.

These mycorrhiza are also involved in anti-biotic and pro-biotic activity[22][23]. There is evidence, for example, that they encourage phosphate solubilising bacteria. There is also evidence that it also seems likely that they may actually discourage or kill some other organisms.

We now know that they actually show significant antibiotic functions.

What the work by Sara Wright does is to show that these mycorrhiza are the source of the protein she has identified as the mechanism which glues the particles together; Glomalin. Now the story begins to hang together and show us what we have to do in the field.

This research has shown that plant root growth is a very large factor in stable soil aggregation, with mycorrhiza closely associated with the process. Anything which encourages root production and mycorrhiza will help form a more stable structure which will produce better crops. Here is the "Catch 22". It is very easy to get onto a declining spiral of loss of structure and pressure on yields and declining organic matter is a major component in that story. On the other hand, if we can reduce cultivations, then organic matter tends to rise and soil structure improves, plants grow more roots, more mycorrhiza grow round those roots, more Glomalin is produced and so on, with a rising spiral of soil fertility and plant growth.

Margins

There is no doubt that soil mycorrhiza are an effective and practical tool in the management of soil structure in the field. So, management of these VAM fungi is a biological technique which can lead to greater financial margins. There are likely to be many factors involved in managing a very complex population which we can't see and of which we have, as yet, limited knowledge. However, there are two very clear factors we can start with and they are closely associated; organic matter and cultivations.

Normally, at least half of a plant's weight is below the surface. So there is plenty of scope to accept that there is at least a significant amount of organic matter in the system. We also know that cultivations oxidise organic matter. So, if the process must cultivate there follows two logical rules; firstly cut cultivations to what is necessary - and only that. Secondly, relate organic matter additions to the violence of the cultivations; if you hit the soil harder with multiple passes of power driven equipment, then higher levels of organic matter additions will be required to replace that which has been oxidised away. "Recreational tillage" (i.e. that which is not really necessary) has an expensive consequence.

Where will arable farmers get the organic matter from? Certainly, conservation of what is produced in the field is the starting point. However, unless the ground is direct drilled or there is some form of significantly reduced tillage, imports will be necessary and an obvious source is waste from municipal authorities and industry. Millions of tonnes of organic wastes are landfilled every year at enormous cost to the rate-payer and the environment. Farming could close the recycling route and give sustainable operation to some of those other businesses. As Dr. Tim Evans, then of Terra Ecosystems (the company which handles London's sewage products), said[46], "We find that many of our customers using biosolids (processed products originating from sewage) against a planned programme report a reduction of crop disease from the start and it builds up." We now know that biosolids, and the new range of products beginning to come from them, have a particular "inoculating" effect with soil micro-organisms and this is another area of soil management which we can begin to manage better.

The pieces of the jigsaw are beginning to come together. We have paid little attention to soil biology in the last 20, even 40, years. Maybe, now we know a bit more about aggregation and how it is related to soil micro-organisms and crop disease, we know enough to make a start on managing that biological balance to margin advantage.

Waste in the Next Millennium

The following piece is extracted from a paper published in "Resource" the journal of the American Society of Agricultural, Biological and Environmental Engineers in July 1998, and is reproduced here by kind permission of the publishers[16]. It does, in fact duplicate some of the information given elsewhere in this book. However, it did two important things at the time which remain important now and, therefore, it is published again here in its original form. Those two things were firstly to crystallise a vision of farmers as a major global force in recycling urban wastes and, secondly, to show that recycling organic materials to land could make a significant contribution to the reduction of crop disease. (The old "crank" point of view of organic faming was, in the respect of crop disease at least, right!)

De-centralised waste management to land offers a major new environmental role for agricultural engineers world-wide. New developments in Europe in spreading direct to land, composting and thermophilic digestion are opening new economic avenues for the recycling of waste. Advantages include recycling as being lower cost than disposal, a significant new source of income for farmers, reduced fertiliser costs, major reductions in rainfall/irrigation requirements, reduced crop protection chemical use and reduced Nitrogen run-off.

What the Rio, Kyoto and Bali "Environment Summits" highlighted was the balance and, indeed, conflict between energy consumption and environmental stability. Consumer societies produce huge quantities of garbage and other wastes. The only factories big enough and sophisticated enough to cope are the Sea and the Land. The sea we cannot yet control adequately but managing the land is Man's oldest technology. In the next ten years, the cost of logistics will push waste handling into an inflation leader. De-centralised waste management, turning waste into fertiliser, doing it *on the farm*, will produce a major contribution to environment management and changing farm incomes. There is indeed here a major new role for agricultural engineers.

Research in the UK is identifying significant advantages in recycling to land. The addition of large quantities of organic wastes or composts has a number of effects on soils. Application rates in the range of 10 to 250 tonnes per hectare produce major physical and biological responses which have economic and environmental advantages. Organic material (in large quantities) will hold moisture, release Nitrogen slowly, change the fungal population of the soil and reduce power used in cultivation.

Back in the early 1990's, EcoSci was a research company in the UK involved in compost research. Table 7.10 shows reduction of crop diseases in plants grown on 100% compost in the laboratory. Such measurements are possible in the laboratory but difficult in the field. However, widespread observations of conditions on farms supplied by the water companies which are responsible for recycling sewage to land confirm that there is a disease control benefit. There appear to be two routes of action. Firstly, dressings of mineral fertiliser such as ammonium nitrate tend to produce flushes of growth which is soft, fleshy and, therefore, more subject to aphid and fungal attack. It also appears that only partly decomposed composts encourage pin moulds (penicillins) and discourage many other organisms. While at least partly understood, it does appear that there are both pro-biotic and anti-biotic effects of using the soil itself as the final stage of the composting process i.e. adding the compost before it is too well rotted. (*In the original research, there was no statement in this paper of why the crop disease was reduced or totally controlled. At the time, the reason was either not known or not clear. As this book shows, we do now know.*)

The reduction of drought stress can be major. As an example, again back in the 1990's, Alwyn Moss was a contractor based at Mildenhall in Suffolk, UK, on the edge of the American Air Force base. The local soil is a blowing sand with frequently less than 380mm of rainfall per annum - the UN definition of desert. Moss took shredded newsprint used as bedding in the racing stables in Newmarket and composted it before adding it to the land at the rate of up to 240 tonnes to the hectare. That newsprint compost will absorb between 6 and 10 times its own weight of water. So, ploughing in the material in the autumn before planting fodder beet in the following spring means that it is possible to grow a useful crop without the irrigation needed by his neighbours. A wider survey of farms in the same area showed a 45% increase in sugar beet yields and an 80% reduction in fertiliser costs for farms using heavy dressings of sewage products when compared with neighbours.

Reductions in the energy used in cultivations following additions of wastes to land are difficult to quantify. Conditions in the field are so variable that scientific statements are clouded or imprecise. However, most workers agree that there is an effect which is, of course, greater where the soil organic matter was previously depleted. Some researchers will offer subjective agreement that these effects may be over 10% reduction in such cases and may be much more significant in seedbed preparation.

When the value of plant nutrients (Table 7.11) are brought into the equation, then one is left wondering why farming is not "farming" waste first and producing crops as a by-product side line.

One of the reasons for slow adoption of waste into farming as part of the system has been easy, "clean" alternatives and not enough economic pressure. Another may have been direct concern about disease and this is certainly likely to be a problem with supermarkets who may wish to control the image of their inputs. There are two developments in the UK which may be of significant value here; Deep Clamp composting and thermophilic digestion.

Deep Clamp composting uses less land than windrowing, does not necessarily need specialist equipment and has a lower surface to volume ratio. How it works is shown in the diagram. (The diagram referred to in that original paper is shown elsewhere in this book at Fig 7.2.) The surface to volume reduction means that pasteurising is likely to be much more thorough. Thermophilic digestion is now being developed with sewage as the fluid carrier and garbage (i.e. MSW - Municipal Solid Waste) in the same stream. New patented methods offer major advantages in single stream waste management logistics with reduced costs and less rejection for disposal tipping to landfill.

The fact is that there is now a momentum in interest in waste into surface soils and new methods to put farming in a position of being waste managers to their nation. This has economic, environmental and political implications which offer opportunities and growth to agricultural engineers.

Table 7.10
Suppression of Crop Diseases by Recycled Organic Material (ROM) Composts

DISEASE	VERY MATURE	MATURE	PAPER WASTE	OTHER
Foot rot of cereals	27%	28%		
Brown foot rot of cereals	62%	86%	37%	
Take-all of cereals	46%	81%	66%	
Blight of peas	NS	66%	61%	Sewage sludge and green waste 49%
Clubroot of brassicas	100%	100%	100%	
Black scurf and stem canker of potatoes				Spent mushroom compost and greenwaste 49%
White rot of onions	75%	90%	67%	

Table 7.11
Fertiliser Value of Composted and Digested Wastes

Guideline figures only. Wastes are highly variable and these figures must be taken as useful guides only, to be modified according to specific circumstances.

Nutrient Figures in kg plant nutrients per tonne of dry material

Original Source	% dry matter	N	P as P205	K as K20	Others
Green waste	40 – 70%	10	10	10	Useful trace elements
Sewage – not digested	4%	50	60	2.5	Useful trace elements
Sewage digested	4%	40	60	2.5	Possible heavy metal pollution
Separated Garbage (MSW) 23 % biodegradable	30 to 60%	60	90	40	Very variable
Liquor from co-digested sewage and garbage (MSW)	10%	50	70	35	Heavy metals very unlikely

Source; OFTec and Bill Butterworth

Plant Metabolism. Anti-biotic Effects - Crop Diseases
Active management of soil mycorrhiza can lead to better crop protection for less money, with a bonus of better yields. Research into the function of mycorrhiza, the fungi which surround plants, indicates that we can develop a line of thought in methods of field procedure to get more control for less cash.

Mycorrhiza are a group of fungi which are closely associated with plant roots. In practice, in the field, that means all plant roots, all of the time. The relationship is symbiotic; both parties get advantage. We know that the relationship is very complex and that it is more active at some times than others. As is always the case, it is not possible to manage with any consistency unless that management is based on at least some knowledge of the mechanisms involved in the process to be managed. We think, from the research so far, that we could, at least in part, make significant progress in managing the relationship. The advantages may be very substantial because the mycorrhiza dictate the plant's relationship with the soil. In effect, they have a very major role to play in water uptake, nutrient uptake, and the whole of the metabolism of the root and, therefore, the plant and the crop as a whole. This will have significant effects on how the plant (and, indeed, the crop) competes with weeds and pests. This, then, is fundamental to crop husbandry and commercial success.

A researcher called Rejon, working for the US Dept. of Agriculture, has worked with a team on mixed plantings of wheat and perennial ryegrass[46]. A trial was carried out on potted plants in the lab under controlled conditions. Selective herbicide was applied to remove the ryegrass. It is possible to produce plants without mycorrhiza on the roots and, in this trial, control was significantly better when mycorrhiza were present. Yield was also higher. The researchers concluded that the mycorrhiza have fungal hyphae which are associated with several plants and, when the weed crop is weakened by the herbicide, these hyphae may be associated with moving nutrients from the region of the weakened plant roots, maybe even from inside the root, to the stronger and un-weakened crop plant.

The questions raised by this research are many. Firstly, it may well be (even likely) that some mycorrhiza species are better than others at helping the effect of herbicides or other crop protection chemicals. Maybe some are better with certain crop weed combinations. Maybe some react better with some crop protection chemicals than others. Turning the questions round, maybe this could explain why some chemicals appear to work better in some situations than others i.e. the right mycorrhiza have to be present. Determining all the correct factors is very complicated but we do have some good leads.

Potatoes and Cereals
EcoSci, based in Exeter, around 20 years ago, did quite a useful amount of quality research on the disease control effects of composts made from different materials. Some of these figures are shown in Table 7.10 and some more in Table 7.12, following, from further trials. Clearly, the mature greenwaste compost had something special. Another laboratory trial by EcoSci gave 100% control of Brown Rot in potatoes grown on organic composts. While, in these trials, there was no examination of the species of mycorrhiza present, the implication could be that the different sources of material and their different conditions affected the activity of the mycorrhiza and how they helped the plant overcome disease. The conclusion that managing the soil organic matter in order to manage the control of diseases better does, in the light of long term, practical experience, look attractive. It must, at least in part, be involved in unravelling the story.

Crop Stress
Research on mycorrhiza shows that they become more effective in assisting their plant hosts as stress levels

rise. If water and nutrient levels are low, then plants grown under conditions sterilised of mycorrhiza are significantly worse off and this effect tends to be progressive; as the stress increases, so does the effect of not having associated mycorrhiza. Again, turning the question round, this is likely to be part of the explanation of why sometimes a crop under stress conditions does better than another for no apparent (previously known) reason.

There is no doubt that mycorrhiza are living organisms and that living organisms need food. Mycorrhiza feed on organic matter. Arable soils with low organic matter can yield well but the technology has to be poured in at ever-increasing levels; mineral fertilisers, crop protection chemicals and mechanical power. Over time, the system progresses closer and closer to hydroponic systems. These systems can work exceptionally well and, there is no doubt, many millions would have starved, world-wide, if these systems had not been developed. However, crop production using high levels of organic matter does have major advantages, especially as costs rise for petroleum-based fertilisers, crop protection chemicals and mechanical power. The best technical results (not necessarily the best financial return) is most likely to come from using high levels of organic matter in the soil *and* high-tech inputs of agrochemicals and power.

There are, of course, many possible sources of suitable organic matter. The best is likely to be the roots of a healthy crop. Such roots will be at least half the total weight of the whole plants and we know that they have the right mycorrhiza attached. Composts made from municipal wastes, paper and other industrial waste will provide good source material provided appropriate

Codes of Practice are followed. One of the best imported sources of organic material will be biosolids. On that subject, we must move away from "sewage" because the unprocessed material has begun to be associated with the possible perception of risk. Modern UK sewage derived products are very underestimated materials and significantly safer, in some cases, than mineral fertilisers which rarely are given a second thought on safety issues. However, remember that digested sludges are different and that thermophilically digested sludge is pasteurised and that puts it on a safety level with milk. The biggest concern about biosolids expressed by regulators is the Heavy Metal content. One of these, to give most concern, is Cadmium. There is more Cadmium in many mineral fertilisers than in a modern UK sewage sludge. Why press the subject of biosolids? Well, it does look as if the bacteria active in the digestion process just might be part of the story on managing soil biology better. Some of the results with biosolids do appear to be associated with mycorrhiza, the right ones, and crop disease control for less chemical. Whatever the original source of organic matter, there is well-established evidence of effect on crop disease.

Table 7.12
Suppression Of Soil-Borne Fungus Diseases Using crops grown in lab in different composts % show level of control

DISEASE	COMPOST MADE FROM:		
	Very Mature Greenwaste	Mature Greenwaste	Paperwaste + Greenwaste
Brown Foot Rot of Cereals	62.3%	86.5%	65.7%
Take-all of cereals	46.5%	80.6%	65.7%

Source – OFTec and Bill Butterworth

Pro-biotic Effects - Nutrient Release and Crop Health
Effect of waste to soil on crop drought stress, plant diseases, nutrient run off reduction, crop drought stress

There is no doubt from the above discussion that organic matter can and does have a major effect on soil mycorrhiza and, through them, very significant and at least partly manageable effects on soils and crops. These effects are all of very significant economic benefit. It is here that "wastes", especially if they contain significant Carbon as large molecules, and preferably a balance of nutrients including all the traces which plant scientists look for, offer a very significant benefit. The condition is that there is a direct relationship between mycorrhiza metabolism and the wastes added. There are, as ever, two components of management which can create that as a successful relationship; the knowledge of the farmer who owns the land and the expert knowledge of the technologist.

Process Capability and Safety
The soil, with the organisms in it, is a staggeringly flexible and tolerant universe. The micro-organisms evolved with plants in an integrated and interrelated package which, on the whole, works rather well. Farmers themselves generally do no go somewhere else at 5pm – if they make a mistake, they know they will hand it on to their children. Add to that an organised discipline based on fundamental science and applied technology and there is the basis of a sustainable operation. The question then arises as to whether that "organised discipline" is the farmers and their organisation, or the state. The logical answer is, of course, a partnership of both. However, in Western democracies, there is a tendency for state regulators to feel safe if they regulate for those who are not responsible, to the detriment of those who are. When responsibility is taken away from people actually on the

ground, either by big business or by the state, then the system will break down and the environment will suffer. Farm-based systems do not have that fundamental fault.

Chapter 8
Food Production

There does not have to be a choice between biofuels and food. Farming can harvest sunlight to produce energy *and* food. There is no reason why a global effort to produce biofuels cannot be accompanied by a parallel global effort to produce food. Indeed, there is every reason why the two should be progressed together as one package. Here's how.

Cutting down the rainforests to grow crops will not stop if the crop is not allowed (by financial or political pressure) to be palm oil for biofuel production. The only route on the horizon at present with the remotest chance of stopping rain forest destruction is to allow, or preferably force, the world's top polluters to buy and protect rainforest in an "offset" scheme. It has been talked about and "progress has been made" but a scheme is not in place to halt all destruction and, frankly, is likely to be a long way off; political wrangling is, as usual, the death of necessary progress.

Case Study
Recycled Wastes into Biofuels & Taking CO_2 Out of the Air
There are three generations of the Bates family farming in the North of Lincolnshire. They have a dairy herd and some beef with 800 acres (about 325ha) of mixed rotation, mainly cereals and oil seed rape. To give a guide, the farm would generally expect winter wheat to yield, average all fields, 3.5 to 4 tonnes per acre (8 to 10tpha).

Brothers, Mick and Philip joined Land Network, the farmer-owned consortium, around 6 years ago with a small scale composting operation building slowly to a current operation of about 10,000 tonnes throughput per annum. That throughput is mainly green wastes (from gardens and collected by the local municipal authority), timber wastes from furniture factories and some waste liquids. (They handle, or will shortly handle, nearly all the materials in Appendix 3. These wastes receive a "gate fee" in the range of £15 to £35 delivered. The farm is building up to fill its permissions for its 75,000 tonnes per annum recycle-to-land facility.

The farm soils benefit from the muck from the livestock but would normally expect to spend around £50,000 pa on purchased mineral fertiliser, mainly Nitrogen. About a third of the farm has now been switched from purchased mineral fertilisers to compost from the wastes. No mineral fertiliser input will be achieved within 3 years. (Note that British farmers import over £1 billion worth of mineral fertilisers pa.)

Both brothers "do what has to be done" as a team, but mostly, Mick drives the composting operation and Philip has driven the biofuel operation. From oil seed rape (OSR), they get 1.5tonnes per acre (over 3.5t per ha). One tonne of OSR seed will give 330 litres of oil and that, after process, will give 330 litres of biodiesel and 65 litres of bioglycerol.

The farm provided biodiesel at the 100% level and to the European Standard, EN14214, for the trials for CNH (Case New Holland) in their world-wide trials of their agricultural diesel engines. Now, all CNH agricultural diesels carry a full warranty for biodiesel up to the 100% level (subject to filter and fuel Standard conditions). The farm has also supplied biodiesel for similar trials in Volvo truck engines.

The point about this case study is that it demonstrated that food and fuel production can go together and it makes good farming sense to do so, it makes good logistics sense to operate both together on a proximity principle basis, and it makes good financial sense to operate a balanced product and business package.

Scaling this up globally will be looked at in Chapter 10.

Human Food Safety
The most difficult question which Land Network has begun to wrestle with is not how to monitor these elements in the soil and in the compost of direct spread waste, it is how to do this economically with the resources which a small commercially-driven organisation can risk committing. The technology and experience does exist but the trained manpower is limited and significant resources are needed. Nevertheless, progress is being made. Safety is not just about "dilution", nor is it just about spreading it out so that nature can cope. It is partly about what has *not* been put on the land. Some arable soils have had no organic manures, no animal waste, no compost, just relatively pure mineral fertilisers, added for over 50 years. Ammonium nitrate is just ammonium nitrate - no trace elements. However, in that time, harvested crops will have removed enormous amounts of trace elements.

This, then, defines the big challenge for recycling to land. We have to replace these trace elements. On evidence, it seems unavoidable that this is a matter of human health and longevity. The most economical way to do this is via "waste", which also appeals as being "natural". We also have to replace organic matter (large Carbon-based molecules). That means that we can use plastics - not robust plastic sheets (such as polyethylene or "polythene") but liquids which can be incorporated and spread out onto every particle in the compost mass. We also need fibres in the soil to manage moisture and gas exchange. There will be less flash flooding if we have fibres to hold the surface open. If the land is subject to intensive cultivation, these fibres may deteriorate (by oxidation) too rapidly and synthetic fibres such as carpets (which have been finely shredded) can and will provide that physical function.

Kinsey and Human Health
Sustainability is the key word; in soils, in cropping and in human life.

Back in the 1950's, Dr William Albrecht had finally concluded how to model the chemistry of farmed soils. He had originally been asked to look at soils which had been broken out of prairie and had yielded remarkably well at first but many went into decline and whatever farmers added in fertiliser, seemed never to recover their yield potential. Albrecht modelled the fertile soils and the declined soils relatively easily. It took years to find a mathematical "bridge" which allowed an operator to calculate how much of what material needed to be added and how long it would take. The bad news was that the calculations were very complex, took a long time and nearly all the world discarded the work as being commercially unusable. However, the fact is that the model not only worked, it showed something else of staggering importance.

In a study tour shadowing the now world-renowned soils expert, Neal Kinsey who took over this work, I was shown a farm where Kinsey had been called in to look at a dairy farm where the grass yields had declined and whatever NPK fertilisers (Nitrogen, Phosphorus and

Potassium) the farm added, the decline progressed. Kinsey modelled the soil and recommended additions of other plant nutrients, including what farmers call "trace elements". It took years but the grass yields did return to "prairie days" fertility. Something else happened; cows "took to the bull" first time, had more live calves and the calves grew faster. The cows had less disease, stayed fertile for more years and lived longer. The situation was complex and, from this one example, it would be dangerous to conclude too much. However, this is not an isolated incident. There is a significant amount of credible evidence that getting the right mineral elements into the ground not only helps the plant (and the crop), but also helps the animal that feeds on those plants. It is entirely logical to progress that thinking to suggest that that food chain does progress right into human health and longevity. There is good evidence to support this[53].

Chapter 9
Population, Deserts and Reforestation
Black Death, Chinese Discovery of the World, Reforestation and Global Warming
Reclaiming Desert
Manufacturing Soils
Using Wastes (MSW) to Reclaim Desert
Water Management
Climate Change - Crops and Irrigation

Black Death, Chinese Discovery of the World, Reforestation and Global Warming

Right back at the beginning of this book, there was reference to the Black Death in Europe and its effect on forestation and global temperatures. The University of Utrecht[31] found reason to link the dramatic reduction of population following a series of pandemics of "the plague", the abandonment of agricultural land, natural reforestation of that land, and changes in global temperatures by around one degree Centigrade. It is worth looking back at the actual dates and population changes, put by many but none more dramatically than William Stanton[4]. Population rose during the Roman occupation of the British Isles up until between 300 and 400AD. There was a fall, due to conflict, disease and slavery, in population between 400 and the arrival of the Normans in 1066 after which the stability allowed population to rise again. That rise continued until the Black Death came out of Asia and, between 1348 and 1400 cut the population of Britain from around 5 million to around 2 million. Now, just hold onto those dates and remember the Utrecht research.

Go back to the work done by Gavin Menzies; "1421 – The year China Discovered the World[3]". Menzies looked in detail at the vision of Emperor Tsu Di who sent his

147

Admirals to map the world and bring it all into China's Tribute System of trading. As Menzies detailed, the Chinese "treasure" fleets did circumnavigate the world and discovered the Americas and Australia long before the Europeans. These fleets were truly enormous with the largest ships over 400 feet long (125 metres) and, the larger fleets over 800 vessels. These were ocean-capable craft and made of Teak. Menzies lists five Admirals, each with a fleet. As a matter of arithmetic, Emperor Tsu Di must have ordered the construction of 3000 to 4000 ships of various sizes, but not small as they were to be ocean-going. The Chinese had also had a merchant fleet for trading in the China seas and Indian Ocean for several hundred years. In the 10 to 15 years before the Admirals set off on their voyages, there must have been construction of maybe 5000 ships. He also built the Imperial Palace and many other buildings – from teak - all from teak. That must have involved the destruction of an enormous number of trees.

Also, some 5000 years ago, the Sahara Desert was covered in trees.

There is no doubt that the loss of tree cover on a global scale is related to global warming and its control.

It is not the purpose of this book to look more at the tree issue other than to observe that the earth used to have rather more of them and they are rather good at taking Carbon dioxide out of the atmosphere and pumping Oxygen back in. So, reclaiming the deserts and planting trees and/or growing food may be an issue worth looking at.

Reclaiming Desert

The Rio, Kyoto, Bali and Copenhagen conferences raised many questions, including:

Is Pushing Back the Desert Worth CASH?

There is a link between current fossilised fuel revenues (or lack of them), desertification, urban waste and biofuels. The real question, then, is how can this link be identified and turned into a business with real jobs, real income and, ultimately for governments, tax revenues.

As the pressure in middle-class Western society, formalised at the Bali and Copenhagen environmental summits, increases its opposition to the chopping down of rain forests to make biofuels, the converse becomes more of an opportunity; to buy biofuels made from crops grown in arid climates and on desert soils. All oil and gas resources have a finite volume; we do not know what that volume is but it is generally accepted that, progressively, they will become more valuable until they actually run out. There is, then, a compelling logic in investing a proportion of oil revenues in reclaiming desert for biofuel production. Again, the converse is also true; if there are no local fossilised fuel reserves, then liquid biofuels from local resources will be increasingly compellingly attractive.

Reducing and reversing desertification may not be as difficult as it might seem. Not, that is, if urban waste is available. Compost made from urban wastes will hold 5 to 10 times its own weight of water. Dressings of 250 to 500 tonnes per hectare, therefore, will hold enough rainfall or irrigation if needed, to grow a crop. Oxidation of the organic matter will vary enormously, up to 100%, but it can be limited to just 10% by cropping, cultivation techniques and waste/compost technology.

PCCS - Photosynthetic Carbon Capture and Storage

Making new oil reserves is quite easy. The dirty black stuff which makes soils black ("DBS" to some soil scientists) is "humus", a complex tar made up of hydrocarbons, carbohydrates and proteins. These molecules come from the degradation products of the micro-organisms which live in soils and compost processes. The feedstock for those micro-organisms is, in natural ecosystems, plant and animal remains. Unfortunately, the whole process in natural ecosystems takes rather a long time – many millions of years. However, just as good, is the organic matter in urban waste. It is helpful, at this point, to go back and look at Chapter 6 and the Figs 6.1, 6.2 and 6.3 which depict the place and function of organic matter in soils and its relationship to crop performance.

The arithmetic is straight forward. In approximate terms for a semi-arid area where there is some rainfall:
- 500 tonnes of "waste" will make around 350 tonnes of compost.
- 350 tonnes of compost per ha will hold around 200 tonnes of water per ha (equivalent to 200mm of rain/irrigation).
- 1ha of crop will probably make 1 tonne of biodiesel and 200 litres of bioglycerol.

To reclaim hot desert, at least double the inputs. In deciding application rates, there needs to be recognition that there is no humus to start off with and the higher temperatures will increase the oxidation rate of the organic matter. This rate will depend on ambient soil temperatures and on cultivation techniques, and these, in turn, will affect crop species choice.

What to plant? Trees give greater year-round stability to soils and the most commonly used palms for biofuels, so far, are Palm and Jatropa. However, there are others. In some areas, Safflower works well and the oil has the advantage that a standard diesel engine will burn it without esterification. Indeed, diesel engine design (and more importantly, engine oil filter development) is progressing and clean oils from a variety of plant species is now possible. In some areas, where there is little or no available urban wastes, the local scrub may be composted on a rotational basis - almost any source of organic matter will do.

How to Push the Desert Back
Reservoirs at the Crop Roots
Trying to grow most crops in areas where the annual rainfall is less than 380mm has its problems; the biggest is that there is not enough water. This book is a new look at old technology in pointing to more successful ways of developing the reservoir at the plant roots. Most arable farmers are conscious of the need to raise soil organic matter but ask where to get it. The use of the waste from the each crop, sewage and municipal and industrial wastes can solve both problems of disposal and of soil water shortage. Accepting other people's waste may bring in very significant revenue.

New developments in dry-land technology suggest that raising the organic matter of sandy soils is likely to be worth 50 to 100mm, maybe 150mm of rainfall per annum - because it keeps the water at the plant roots rather than draining or evaporating away. There are other nutrient and disease advantages. These figures have come from farm studies, almost unnoticed by formal research, and from all over the world on the disposal of waste (sewage and municipal waste) to agricultural land.

Growing crops successfully in any dry area or year does, of course, depend on many factors. For centuries, there has been much thought and effort put into irrigation. That, however, is expensive in resources, will always have at least some salination effect and there may not be enough water anyway. Extraction licenses, to take water from rivers and boreholes, are often not easy to obtain from regulating authorities. There has, however, been comparatively little attention to the reservoir at the plant roots.

The most obvious limit to the soil reservoir is the depth to which the plant roots penetrate the soil. Many vegetable crops will put their main roots down near two metres in one growing season. So deep cracking of the soil to break pans and compaction is commonly practised - but not always when it should. There is a less obvious factor which is commonly neglected - the soil organic matter. The water retaining capacity of the soil depends on a number of factors including the mineral nature of the soil itself and the soil voids which are affected by cultivation. However, soil organic matter has a major effect and it can be raised by thoughtful management, or dramatically reduced by cultivation.

Low organic matter soils are much less tolerant of heavy machinery and compact more easily. Further, it is the organic matter which is the main "reservoir" for water.

There is a problem in cultivation; aeration will dramatically increase oxidation of organic matter. This rate of oxidation is faster and more destructive in hot soils and climates. So, in cultivation, a balance has to be struck; the higher the temperature and the more thorough the cultivation, the more rapid the destruction of organic matter.

Higher Yield

In one study in the East of England, several farms involved in a wider scientific study were noticed to have abnormally high yields of sugar beet and other crops. Three of the farms, which were on Suffolk sands, yielded consistently higher than neighbours, all of which had apparently similar levels of husbandry. The high yielding farms had been using sewage sludge for between 5 and 30 years as their main source of fertiliser and achieved yields for sugar beet typically of 16 tonnes to the hectare compared with neighbours at 11 tonnes. That was a 45% increase in yield accompanied by a 75% reduction in expenditure on mineral fertiliser.

Compost

There is increasing evidence of extra disease resistance of crops grown on soils fertilised with compost made from organic matter. Some work done by EcoSci Ltd of Devon in the UK, has shown that crops grown in the laboratory on green waste compost showed remarkable resistance to disease. There was 100% control of Brown Rot in potatoes, 100% control of Club Root in Brassicas and typically 65% control of several cereal diseases.

Alwyn Moss, farming near Mildenhall in Suffolk UK, has composted tens of thousands of tonnes of shredded newsprint (used as bedding for horses at the Newmarket racing stables) and applied 250 tonnes to the hectare - that takes some skill to plough in. The compost absorbed 5 to 10 times its own weight of water. Ploughing in before the winter rain will provide a reservoir to produce a crop without irrigation. That means that this dressing of compost per ha will absorb and hold up to 2500 tonnes of water per hectare, equivalent to 250mm of irrigation.

There is another potential advantage of some economic and environmental significance. Less run-off of mineral Nitrogen into water courses will make a useful contribution to reducing water pollution.

Sources
Even in desert areas, there are potentially four sources of organic matter which may be within striking distance and relatively un-tapped; the waste from the last crop, sewage, industrial wastes and separated or whole MSW (Municipal Solid Waste or "bin rubbish"). In some areas, **unseparated** MSW may be the (perhaps surprising) opportunity.

Raw sewage has many problems if it were to be used on crops grown for human consumption. In most countries of the world, however, filtered sewage is commonly so used. Most sewage digestion in the UK is mesophylic (carried out at up to about 38° Celsius). New technology in thermophylic digestion (at 60° Celsius) will provide major advantages. Firstly, it pasteurises the waste which will kill most pathogens. Secondly, it will allow co-digestion of added solids including crop waste or whole MSW. Thermophilic co-digestion yields much more gas (which can be used to drive engines and/or generate electricity). The liquor has at least 10% dry matter and is an excellent and safe fertiliser and, of course, it is mainly water. In the best possible situations, sewage from a small town or village can be piped to an on-farm digester, eliminating all trucking.

Inevitably, not all situations fit these ideal solutions and there are various levels of sophistication and engineering built into other approaches to utilisation on the land.

One of the counter-balances to rising population and industrialisation is the increase in waste. As wealth rises, so does waste. Perversely, as the waste goes up, so does the opportunity to use that waste, via composting, to produce crops without mineral fertilisers and to reclaim land for crop production with the green leaf taking Carbon dioxide out of the atmosphere and pumping Oxygen back in.

OECD (Organisation for Economic Co-Operation and Development) figures indicate that one person's output of waste per annum is;

USA	750 kg per person, per annum	
UK	580	
Japan	400	
China	120	
India	100	

These figures give some idea as to the opportunity to use wastes for land reclamation, for making fertilisers and growing crops.

Manufacturing Soils from Wastes
It is possible to make a soil from "wastes" which will, over time, be at least satisfactory, and possibly really good. There are three basic fundamentals: the physical components and structure, the chemical constituents and the biological activity.

The physical structure of soils involves mineral particles (sand, silt and clay) in any proportion and containing stones and all sorts of "foreign" bodies. This structure needs to present an anchorage into which roots can (reasonably easily) grow. In uncultivated soils, this may, in time, become rather like a stack of uneven concrete blocks with the gas exchange, moisture movement and root penetration mainly in the place where the mortar would be in a wall. If these movements are restricted, then it does not work so well. In a heavily cultivated soil, this structure will break down and future cultivations will need more power, the crop will need more nutrients and each plant may become more susceptible to disease. High-tech farming can go a long way to living with these problems.

A chemical structure which provides a balance of crop nutrients is the second essential. The absence of enough of the right nutrients in the right proportions to each other will result in a crop limited in its growth and more susceptible to stress from disease, drought, water-logging and wildlife attack. The range and balance of what these nutrients might be can be studied in many other texts. What this text has concentrated on is the contribution to crop nutrition by composts and refers to the work done by Albrecht and Kinsey[20][28]. It is not just the nutrients as a list, nor just the individual supply of each nutrient, but also the balance of these elements and how this is related to the physical structure of the soil and the presence of humus.

The biological activity in the soils is equally important as the physical and chemical status of the soil. The evidence is that crops are fed substantially through the mycorrhiza, certainly so in natural ecosystems, or compost-based crop production. In mineral fertiliser-based crop cultivation, it is likely that soluble materials usually enter the crop roots directly, avoiding the mycorrhiza at least to some extent. The evidence in this text points to greater stability to the farmer and a move on all fronts towards true sustainability, by moving to humus/mycorrhiza-based soil management. This means that the mycorrhiza need feeding and, fortunately, they can be managed much as crops are managed with mineral fertilisers, except that the nutrients need to come via humus and Carbon chain molecules.

So, in temperate climates, we can take construction waste "fines" (the sandy materials passing through a screen) with the right chemical composition, add a suitable compost (again with the right chemical composition) and produce a potentially productive soil. Suppose we apply this thought to reclaiming erosion-depleted upland soils or to making the deserts of the arid areas of the world productive - as many used to be? As an example, Malta is a relatively small island with a maximum distance, coast to coast of only 7 times its airport runway length. Malta has around 400,000 total resident population and around 2 million tourists a year. Despite this population pressure, over 70% of the nation's waste output is construction waste. The island was almost certainly covered in much more vegetation historically and yet much of the surface of the island, that which is not covered in concrete, is eroded back to the bare rock or, at best, very thin soils except in a limited amount of agricultural soils (which are still quite shallow and mostly poor in organic matter.) That bare rock could be recovered back to vegetation and trees by using crushed and screened (or just screened) construction waste plus composted organic wastes and biosolids. (The large particles from screening could be used for wind-protection bunds, inland dams or sea walls.)

The same approach could be used in the UK to reclaim uplands denuded of soil by over-grazing in the past; similarly in most of the countries of the developed world.

Using Wastes (MSW) to Reclaim Desert

To make an arid soil productive, there needs to be nutrients and stability in the face of erosion – particularly wind erosion in dry times and water in a rainy season (if there is one). Wastes can be used to do that. The safety framework needed is control by agricultural scientists with local knowledge, analysis of incoming materials and the use of an Albrecht-Kinsey soil monitoring programme. With this framework arid soils can be managed to produce tree crops (including oil-producing Jatropha and oil palm) and food which will contain a good range of trace elements and will probably help people live longer with less disease.

Waste has always been a political issue in every country in the world, but it is comparatively recently that *recycling* of waste has taken top of the waste agenda and the word "recycling" has come to mean high-tech solutions and, very often, Energy from Waste (EfW). However, waste can be used to recover arid and desert soils, make them productive and, in so doing, remove enormous amounts of Carbon dioxide out of the atmosphere and put Oxygen back in.

Whole, unseparated MSW (Municipal Solid Waste) can be composted to produce a safe material which can be used as a nutrient-rich soil stabiliser[49]. It may be safer and more economic to remove "tin" cans and glass but they can be left in and the following system will still work. Cans can be reasonably easily removed by magnets which will remove the Zinc used as a coating and the ferrous metal may be worth net cash. (Zinc is important as a trace element in the healing mechanism in human metabolism but can be toxic if too much is present.) Glass may be dangerous simply because it may cause injury by cutting flesh. Despite these dangers, with adjusted management, it may be possible that both can be left in safely.

The best method of composting is to pile up in a large heap, around 3m deep and turn across a screen. What goes through the screen can probably be used as a compost for fertilising, quite safely, food crops. What goes over the top of the screen can be used for stabilising the top soil. It will look untidy but it will do the stabilising job.

Trench Reservoirs
In extreme arid conditions, if a trench is cut, at least half a metre deep, filled with compost and then covered with a mulch of larger particles, just mixed into the surface, say, 50 to 100mm deep, with the trees planted though the surface layer, then that will provide a reasonable start with minimum irrigation need. What the material in the trench does is to provide a top-soil reservoir; most composts will absorb and hold 5 to 10 times their own weight of water. What the top layer does is protect the start-up from wind erosion and limit the damage from sunlight oxidising the organic matter in the trench reservoir. An alternative top layer is one of sheet plastic (such as polyethylene – which is commonly known as "polythene").

The nutrient value of the compost will, of course, vary with the input material. The stability of the surface mulch and how untidy it looks will, in turn, also depend on input material but also how it is treated. According to Ashish Kumar Singh[36], working with colleagues in India, MSW will contain over 75% biodegradable material and, in that, there is plenty of nitrates, Potassium, Magnesium and other plant nutrients.

159

In the developing countries, the largest proportion of MSW is what the research describes as "biodegradable". However, the truth is that everything is biodegradable but some materials are more easily degraded, or faster, than others. It is, in the context of this discussion, useful to distinguish between easily (such as food waste) and slowly (such as plastic containers) degradable materials. The material which goes through the screen in the process described here can be safely used for growing food and the material going over the top of the screen can be put down the trench in the tree-growing technique. The slow-to-degrade material is exactly what is needed to stabilise the surface layer. If it has particle sizes limited by passing, say, a 100mm screen, then it is easier to handle and mix with the topsoil and looks less untidy.

There is good sense in the old adage of "keep it simple" and there is no reason why whole garbage cannot be recycled in this way. It would be sensible to analyse the various grades of compost and watch, in particular, heavy metal build up. There is an alternative in that the "Smart Truck" briefly discussed in Chapter 7 has some special advantages in developing economies;

i.   It allows maximum retrieval of materials with immediate cash value.
ii. It provides a route to incentivise recycling.
iii. It is flexible, the incentive can be cashed in supermarkets, breweries, the church school or even the tax authorities.
iv. The good grade compost can be used for food crop production and is able to retain moisture and reduce irrigation need.
v.   The coarse grade compost produced can be used to establish energy crops such as Jatropha, again with less irrigation need.

Is recycling "wastes" to land dangerous? Potentially, yes, there are significant dangers of long term pollution of soils. However, it is important to put the risks into context.

Zinc is one of those "heavy metals" in garbage and also in sewage and biosolids where it comes mainly from what women put on their babies' bottoms (for protection against "nappy rashes") and on their own faces (in cosmetics). Zinc is also one of the trace elements essential in the healing process in the human body and is often added to animal feed in a mineral and vitamin mix. So an informed balance needs to be kept.

The absence of concentrated industrial areas, as is often the case in developing economies, may reduce the risk of heavy metal pollutants in garbage. Where light industry is integrated into domestic areas may change the risk: in this case, heavy metal monitoring is likely to be more important. In any case, we have the technology to process urban and many industrial wastes safely. We have the technology to plan what should be put on the soil, monitor the soil and protect it long term. Give farmers the technological support and the job can be done. There is one further area of input which is likely to be wise and high-tech. It is the long term monitoring of land if it is to be used for food production. The living universe of a productive and biologically active soil is continually changing its nutrient balance and is quite capable of locking up or dumping elements into the groundwater. Too much Sodium can, for example, be pushed out of the soil by adding a calculated quantity of Sulphur in the right form. This technology was initially developed by Dr William Albrecht and later by Neal Kinsey. It is now more widely used but is far from an everyday practice by amateurs. Nevertheless, it does give the safety management which is needed long term if wastes are to be the main source of nutrient supply.

Summary
There is no doubt that wastes can be used, sensibly and safely to grow food and energy crops. Imported energy is expensive. Waste disposal is expensive, unproductive and generates future problems. Recycling wastes to land to grow food and energy crops makes environmental and financial sense; it is very attractive. The bonus is that planting trees, preferably for energy production, on recovered land takes net, enormous amounts of Carbon dioxide out of the atmosphere and pumps Oxygen back in. The fact is that the whole concept is cash driven by the cash involved in handling "wastes".

Climate Change - Crops and Irrigation
Whatever you believe about global warming, it will bring major, fundamental changes in the way agriculture is run and, in particular, the way water is managed. So how abrupt will the rate of climate change turn out to be and what can be done to plan farm production to accommodate and exploit these changes?

Few who talk about climate change will know much about, or have even heard of, "global dimming". When fossil fuels are burned they produce Carbon dioxide, a "greenhouse gas". Increasing amounts in our atmosphere allow more sunlight through to the surface of the earth, and air and surface temperatures rise. Most observers, including those in the USA, accept that this is happening although there is discussion about how fast. Clearly, whatever the speed of temperature rise, it will affect agriculture. It is doing so already. Global dimming counter-acts that effect. One of the parallel effects of burning fossil fuels and of industrial activity is that dust, soot and "dirty" chemicals are produced into the air. These have many effects, most of which are seen as harmful. However, it is also true that they tend to settle and concentrate on clouds in the sky, thus turning the clouds into mirrors which reflect sunlight back into space. This reflection of sunlight by clouds is called "global dimming". So pollution has, to some extent, been balancing out with the dimming effect of dirty clouds, thus compensating for greenhouse gasses. How good is this balance? Well, one of the more interesting results of the 9-11 terrorist attacks in the USA is that most aircraft in the world stopped flying for 3 days. Some research studied the climate effects of this loss of aircraft burning fuels and its effect on global warming. There was some evidence that the environmental legislation driving at cleaning up our air could dramatically reduce global dimming and temperatures across the globe could rise by 10°C within a short period of 10 to 20 years.

In actual fact, not enough is known to be sure of the rate of rise in temperatures, but one of the urgent consequences is the need to plan water management.

The Opportunity and The Answer

Water, for many, especially in warmer areas of the globe, is likely to become as important as petroleum fuels. If you have water and can manage it in the face of whatever these climate changes bring, then you will be better placed than most in maintaining life and growing business.

The building of better water management is essential at every level, including international and local politics, industrial planning, and at farm level. This means managing the "water supply chain". Our society's survival depends on our ability to manage water better and better. We need to examine losses and efficiency of use at every step. Perhaps this is no more evident than in Israel where high-tech farming produces good crops from recovered desert but there is a threat of water running out; the "water clock" in Israel is ticking.

A simple check-list for immediate action might be:

Aquifers: Where are they? Can we find more? What are the risks involved in their long-term use? How can they be filled and used efficiently? How can their pollution and long-term damage be avoided?

Dams and Reservoirs: Do we need to build more? When? Can they be built at farm level? Can dam building be integrated with energy production from hydro-electric schemes?

Water Distribution: Distribution can, and often does, entail losses of up to 50% of what was conserved in the reservoirs. What exactly are these distribution losses? Can they be eliminated, reduced or exploited for some use?

Air Management – Tree Belts: Tree belts can change local climate and reduce soil erosion risks.

Top Soil Reservoirs: Raising soil organic matter can increase its retention rate to hundreds, sometimes thousands, of tonnes of water per hectare from the rainy season into crop growth and harvest. Cultivation oxidises that organic matter. Raising organic matter can easily be done using urban wastes[48].

Irrigation: Surface irrigation using hand-dug channels is cheap but extremely wasteful in water use. At the other extreme, a "spaghetti" line to each plant is efficient but expensive. Climate change will push the balance in the direction of more expensive systems in order to use water better.

Cultivation Management and Zero Tillage: Cultivations turn moist soil upwards to be dried in the sun; this oxidises the organic matter. Zero tillage systems dramatically reduce these effects. So cultivations and cropping may have to change to take advantage of these facts.

Crop Choice: Some crops can cope with drought stress better than others. Similarly, some varieties within a crop species are better in this respect than others.

Water as a Political Issue: We can view water as maybe *the* most important of all our assets which we must manage better, using all our existing knowledge plus a major research effort to do better.

Chapter 10
Scaling It Up To Be Global
Anywhere in the world
Building a community-based landbank
Land Network as an organisation
Economics of the recycling to land as a business
Costings of inputs of men, machines and professional services
Outline business plans

Building a Community-Based Landbank

Land Network as an organisation was conceived in the UK back in 1992 when a farm contractor, Tony Birchnall, went to the Marketing Initiative of the DTI (Department of Trade and Industry of Her Majesty's Government in the UK), and said, "I see the writing on the wall for my business, I'd like to look at waste recycling. Can you help?" The project arrived on the desk of the author of this text. What emerged as a project was staggering in its clarity:

"How could the resources which already existed in the rural economy be harnessed to recycle to land safely and economically and so as to avoid the purchases of imported mineral fertiliser?"

There were two important parameters required to build an organisation;
• a technology and legally-based discipline which earned the respect of all the stakeholders, and
• a recognisable respect for the sovereignty of individual farmers.

In both of these, the word "respect" occurs. It might also be the word used in terms of respecting nature and its processes in order to create safety and sustainability.

It is sometimes said that no-one has ever made British farmers co-operate on a production operation. Whether that be true or not, the difficulty would be recognised world-wide. The defence-beyond-reason of sovereignty of landownership or occupancy is instinctive in almost all farmers, everywhere. Therefore, in this project, it was necessary to avoid any question of compromising the independence of the individual. Instead, something had to be added and obviously so. The model came from JC Bamford, the man who developed what might technically be referred to as a back-hoe excavator but is commonly known, world wide, as a "JCB". In order to sell his machines, JCB, the man, gave machinery dealers a "franchise" but insisted they set up not just a separate department but a separate company and that company would include in its name the letters JCB. He also took a 10% share in that company and sat on their Board. So, he knew what was going on.

Land Network was set up in a similar way but with a twist. The farmers were offered two important safeguards. Firstly, none of the trading was done through the farm accounts. All of it was initially done through the central support operation which was set up as a separate company. So, if anything ever went wrong, it did not affect the farm itself. The central company had no assets and took legal responsibility for Compliance with the Law and on Health and Safety matters. That meant, for example, that if what became known as Central Support, led by the General Secretary of Land Network, wrote a Code of Practice, and the farm stayed inside that code and there was an accident with injury to a human, it was the General Secretary, not the farmer, who was morally and legally responsible and risked going to prison under Health and Safety regulations. These safety mechanisms were fundamental. Indeed, the whole ethic of the organisation was, and remains, risk identification, evaluation, isolation, management and all those functions which go into risk analysis and management. For that risk management to be successful in the short and long terms, there clearly had to be a discipline which was imposable and sustainable. In many circumstances, risk management can be imposed and, with willing participants, can be continually developed. That continuation necessarily needs that willingness.

What was the twist to add to the JCB model? Well, farmers could progress to have a share in the organisation and they could actually own it. The design was complete. It did provide a start which was acceptable to some, even many. As individual farm companies grow, they get to own 90% of their own Land Network company with the Central Support unit in the centre owning the remaining 10%. It is still developing 15 years later, but that basic structure remains; it is a franchise where the franchisees get to own the franchisor in what became known as a "reverse franchise".

This approach to community-based organisation could be applied anywhere in the world, not just with farmers but also with any trade – plumbers, bricklayers, cleaners, accountants and anything in suits. It takes nothing from the individual except the compliance with a discipline which is, in any case, a necessity in modern society. In return, it gives a support which could not be afforded by a small individual operator and a brand image which can develop into something with significant economic and political clout.

Land Network as an Organisation

As a national, farmer-owned consortium, the function of Central Support in Land Network is operated by Land Network International Ltd which started out as what many would call a consultancy company; the one which worked for the DTI at the start of the project. That company is also being structured so that it is progressively owned by its staff. Central Support provides Codes of Practice based on common sense, farmer-members' own knowledge, technology, regulation and, perhaps most important of all, the developing pool of experience which the organisation has as a whole, i.e. all of its Members. The central operation also provides supervision, advice, commercial assistance and whatever support members need to develop their businesses.

Initially, the farms do all the physical work and Central Support does virtually all the paperwork. The farms are supplied with the necessary paperwork, including data sheets for recording waste "in" (weighbridge ticket or other means of estimating load quantity and identity) and support to manage their own bank account. Central Support collects the Gate Fees and takes an agreed, small percentage, before passing on the majority of the cash immediately on receipt from the waste hauliers. If and when an individual farm grows big enough and if all the conditions are right for sustaining that growth, then they are set up with their own Land Network company which carries a regional label. Examples of real commercial operations working within the framework of a national consortium can be viewed further at www.landnetwork.co.uk. The following figure shows the basic structure of Land Network as a reverse franchise. The farms remain as individual, sovereign businesses but they may operate in a group in order to serve one large supply contract.

Figure 10.1

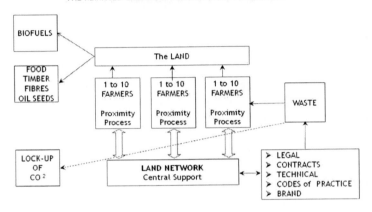

Figure 10.1
THE MANAGED LANDBANK USING REVERSE FRANCHISING

Central Support is responsible for disciplining the supply chain which starts with "waste" at the point of its creation and ends with the production of crop and animal products. Increasingly, using the land as a Carbon sink to lock up Carbon dioxide, is seen as a product.

Fig 10.2 shows part of the Land Network consortium in the UK in 2009. This is a development related to investment in much larger sites with significant investment in processing facilities. The structure is the classic upside down pyramid, starting with the Central Support Directors who support the regional companies. They then, in turn, operate their own industrial composting facility and their own landbank. This structure was developed from the original basic Land Network reverse franchise (it is in fact almost identical) but, in this case, it was designed to develop major industrial processing facilities, capable of composting in-vessel and in tonnages ranging from 25,000 tonnes per annum up to 125,000 tonnes pa in one site.

Figure 10.2

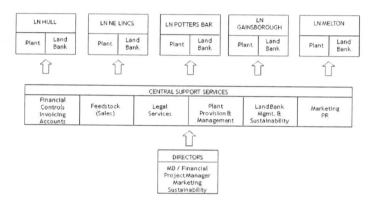

Figure 10.2
THE LAND NETWORK 'DUST TO DUST' PROGRAMME

Economics of Recycling-to-Land as a Business

Until the end of the Second World War, the mass production of low cost mineral fertilisers and the logistics of global distribution limited the impact of these potentially highly productive fertilisers on world food production. In the 1950's, however, the impact was staggering. When mineral fertilisers were low cost, up to around the late 1990's, the developed farming systems of western nations found it convenient, productive and profitable to use them; to the point of not bothering with recycling of wastes. The USA developed the "Land Bank" programme to pay farmers not to produce and Europe even invented "Set Aside" to limit production.

Both of these, to the shame of the countries concerned, occurred when over half of the world was short of food.

The global rises in fossilised fuel energy costs in the 1990's onwards have made that attractiveness less and less profitable. The scope for recycling "wastes", however, has become dramatic. Proximity recycling gets trucks off the road. In the UK, British farmers in the first decade of the 2000's were spending around £1 billion on mainly imported mineral fertilisers. That import bill and its effect on the country's balance of payments could be wiped out by using recycled waste compost as mineral fertiliser.

Back in the early 90's, the Enterprise Initiative of the DTI funded a series of studies looking at recycling urban wastes to land. Some of these studies ran costs into eight figures but there was one which ran a total under £40,000 before the progress generated began to become self-funding - it resulted in the farmer-owner consortium "Land Network" which, between the individual Members, has, at some point, recycled all the materials in Appendix 3; successfully, safely and within the regulations in force at the time.

Part of that original study looked at how much "waste" there might be nationally which could be recycled to land sustainably. The figures were potentially unreliable but, after many discussions, including with what was then the Centre of Waste and Pollution Research at the University of Hull, the study concluded that there was possibly 100 million tonnes per annum. Land Network now concludes that the figure is higher, potentially much higher. The total land area of the UK is just over 24 million ha but less than 20% is arable and just over 50% grassland (productive grass, of course, requiring, high Nitrogen fertiliser input). Forestry would be more productive if compost were applied, too. Therefore, around 10 million hectares could be used for compost substitution for mineral fertilisers. At 25 tonnes per hectare of compost, bearing in mind that composting loses maybe a quarter of its weight, that means that the available land could use in the region of, say 30 tpha of feedstock and a total of, possibly, 300 million tonnes. We have enough land.

What will make this happen? What is the driver? Gate fees for the individual farmer; gate fees which are driven up by regulations. What drives up the gate fees? Regulation! To that extent, the EU and UK governments got it right! All these regulations really do is to provide a framework to organise recycling of "wastes" which society has to spend resources and money on anyway.

Costing of Inputs of Men, Machines and Professional Services
Anybody with business skills can take local costings and write a business plan for a waste-to-land recycling operation, whether it be via composting or direct spreading. If the gate fee for each material to be taken in is known, then there may be some security and value in that plan. However, with the present rapid and significant changes in regulation, technology and energy costs, long term planning is not easy unless there is a matching long term plan of supply and gate fees. This is why companies offering Energy from Waste (EfW) and Mechanical and Biological Treatment (MBT) plants, which may cost several hundred million pounds Sterling, often demand 20 or 30 year contracts. The problem with such a commitment is, again, the rapidly changing business and technical environment in which we operate. It is inevitably true that these plants will

become superseded and out of date before or shortly after they are commissioned, let alone before they complete their design and financial investment life. The reason for these huge EfW and MBT plants is, too often, because of political and misguided views or "advantages of scale" and statutory targets, rather than because they offer the best long term solution both financially and environmentally. The need for flexibility in these big facilities is not easy to satisfy and it is not the main subject of this book. However, there is one real advantage of on-farm recycling to land; it is relatively flexible.

Outline Business Plans
In the planning of an on-farm composting facility, the unknowns can, to some extent, be examined and planned for with sensitivity analysis. Sensitivity analysis (in this case relating a range of gate fees to a range of tonnages) can be related to capital and running costs. But in addition to this, it is the vision of waste integrated to its value on the land which has driven the growth and success of the Land Network farms. That is developing into sustainability not only in the narrow farm/environmental sense but into the global sense including energy sustainability. More information related to individual circumstances can be obtained via the Land Network website at www.landnetwork.co.uk

Bali and Copenhagen
The United Nations environmental summit at Bali in December 2007 got the ball rolling, with the promise that Copenhagen in Mach 2009 would actually give new and real direction in control of global warming and in the reduction of greenhouse gas production. What happened in Copenhagen was that, yet again, scientists urged governments to accept that the speed of progress

174

of global warming was greater than previously accepted by consensus. In these discussions and changes in attitudes, there is a potentially enormous cash possibility in the form of Carbon Credits. Basically, Carbon Credits allow someone who can prove that they are reducing Carbon dioxide in the atmosphere, to sell that evidence of Carbon capture to someone, such as an industry burning fuels, who is producing Carbon dioxide. The idea is to balance the two. It may be that owners of forest can offset their Carbon capture capability, in a formal way, to obtain Carbon Credits with a substantial cash value. The question is what will be defined as "forest"? Could it be that the definition will include palm plantations which will be fertilised with compost made from "wastes" and planted with trees that produce liquid biofuels? There is a certain logic about this and there is only limited scope for farmers to state this case and get it onto the agenda. There is a case for each of us as individuals, each nation, to stop, think, and act now.

Summary
The fact is that we really do not know at what speed global warming is occurring but there is little doubt that it really is happening and at a speed which is disturbing. Anyone who does nothing is certainly letting their children down and probably themselves. Yet the opportunity still exists to consider the options, develop new cropping and new business opportunities.

It is the use of business that will drive this. It is only the business manager who has the skills to drive this and the opportunity is there. It is governments which can open that route and provide the framework of safety.

Chapter 11
Bureaucracy - Politics and its Civil Service

To be help or hindrance; that is the question. (Apologies to Shakespeare.)

Jim Collins, in his book[51] on his excellent research into what makes a really great operation, made a very interesting and succinct observation about bureaucracy. The research team looked at, amongst other operations, that of George Rathmann, the co-founder of Amgen, a very successful pharmaceutical company in the USA. Collins saw that Rathmann understood that "the purpose of bureaucracy is to compensate for incompetence and lack of discipline". In that, the regulators clearly have a function and the State, wherever it is, has a duty to set up just such a control mechanism; there is always incompetence and indiscipline in every area of society and nowhere more than in handling wastes. However, Collins goes on to observe that what happens in most organisations is that the culture of the bureaucracy is to develop itself in order to build increasing improvements in risk management and safety levels. This is encouraged and accelerated by the small percentage of operators who are prepared to take unacceptable environmental risks and continue, in their own culture, to try to get round the regulations and the regulators. This tends to increasingly inhibit the disciplined operators who really want to "do it right" and they get faced with a choice of bending the rules or going out of business.

The alternative to such a negative development is that the regulator has to find a way of imposing a discipline without switching off the good guys. This is always possible if the right minds are put to work on the problem and the political will is there. There are potentially many solutions to this and one is outlined below.

There is one more pitfall to be recognised before looking at the potential solution. There is a cancer-like mechanism in bureaucracy. It is a bit like coat hangers; put them in the dark and they breed. The democracies of the Western nations have a real problem and it is the growth of numbers employed in the state sector. Generally, two regulators do half as much effective policing and twice as much inhibition to good operators, as one regulator.

Most economists agree that when public expenditure (of a state) gets to 50% of total national spend (as GDP), the situation is, at best, potentially very difficult. Some economists argue that at that point, total collapse of the economy is unavoidable.

The historian Jane Marshall once said; "It is in the history of the world that, whenever an empire collapses and for whatever reason, those left in power in the middle pass more and more regulations, involving more and more public servants, in order (they think) to reverse the collapse. What actually happens is that they stifle innovation and inhibit entrepreneurial activity, so accelerating the rate of decline. That is what is happening in the UK, here and now."

Running an efficient and enabling civil service depends on having a very clear vision of what the objectives are and a very clear discipline of the joint obligations of policing and enabling. Generally, that cannot be achieved with more staffing, it is achieved with less staff, properly trained, properly paid and properly led. This comes down to leadership at the very top, clearly

177

understood mission statements and backing of the people. Bringing this back to reversing global warming is a very difficult task for any leader.

Even slowing down global warming, never mind stopping it or reversing the process, cannot happen anywhere in the world without political decisions backed by political will on all fronts, to make it happen. Politicians are frightened that their country might lose out and they personally would be seen to be responsible. To take these risks on is a very tall order; countries are unlikely to develop a consensus for action which is active enough, until it is too late. However, there is nothing like pure, undiluted greed for driving a situation. What it comes back to is that environmental sustainability cannot be achieved without financial sustainability. If the cash is there, then the process can be driven forward against the clock. The only real hope is that there is both political will *and* entrepreneurial opportunity.

It won't happen anywhere unless it is financially sustainable. The solution posed here is financially sustainable and has been done. (See the Case Study in Chapter 8.) It is a workable, in-place solution with a practical track record. It needs scaling up but it does work.

Would it work if it were scaled up? Could it be scaled up far enough to crack a global issue of this nature? Certainly not by some super, patented gadget to be made in factories or by burying Carbon dioxide in porous rock a couple of miles down. However, yes, *technically*, it really can be done. It has been done before in the Carboniferous Era. We *know* it works.

The Structure: Can it be scaled up? Well, Chapter 10 gave a practical solution involving a mechanism to manage a landbank which can recycle wastes to land. The "reverse franchise" mechanism developed in Land Network does actually work. Farmers in it do co-operate and they do include in that a high degree of self-regulation.

The Permit - "Driving Licence" Permits: Land Network has suggested that a new system of Permits would dramatically simplify the process of application and policing. The development would change the nature of the Permit (and all the Exemptions) to a format similar to the driving licence used for car drivers in the UK (which, incidentally, has the safest roads in Europe and possibly the world). Waste management facilities would get a license easily and quickly. This could be controlled by either the responsible operator passing a test or examination or having an established track record of safety and compliance with the law. Permissions could be progressive, with record earning wider permissions. Random inspections would establish if a set of simplified regulations were being adhered to and, if not, a penalty of, say, 3 points added to the site record. Cause pollution and the record gets, say, 6 or 12 points depending on significance of pollution. Get 12 points and the gate is closed to any new business for, say 1 week in the first instance, 3 months for a repeated offence.

One more idea; to provide a global mechanism to manage the emissions from aircraft, set up the United Nations to levy and collect an aviation fuel tax. There are comparatively few oil producing countries and companies and the mechanism could be manageable. It could also provide a model to build on.

Chapter 12
Priorities - Making it Stick

Global Warming
The inescapable conclusion of the research is that global warming is happening and probably faster than we care to admit. We don't really know how fast but it is logical that it would be prudent to get on with doing what we can.

In one sense, it does not matter whether the change is natural cycles or man's folly. If the current growth in global warming is *not* entirely due to man's folly with unfettered population growth and burning fossilised fuels, then we will certainly still need the skills and commitment to applying a reversing mechanism so that we can keep both human folly and natural trends within limits.

Tree Issues
These three basic issues; limiting population growth, stopping burning fossilised fuels and switching to biofuels all need to be done as a matter of urgency. Each must, therefore be started and run parallel. Politically, there needs to be a series of steps with very clear messages.

Population
The first is to find a mechanism to stop population growth. A not unrelated issue of creating some sort of stability in the status quo is to stop the destruction of rain forests – Carbon offset is a real possibility. (On this same subject of destruction of the rainforests, this has nothing whatsoever to do with biofuel production; palm oil is just one of a number of cash crops which drive the destruction. It is cash which is the driver and, if there is

cash involved in their preservation, then they will be preserved.

Fossilised Fuels and Biofuels
The second issue, dependence on internal combustion engines, can also be helped politically by opening a route to profitable biofuels production; part of that will be to tax fossilised fuel use and to allow the United Nations to tax aviation fuel.

Land Use
Finally, reforestation and the production of biofuels from crops grown with compost made from wastes, rather than using mineral fertilisers, is a real way to stop and reverse global warming by taking Carbon dioxide out of the atmosphere and pumping Oxygen back in.

CASE STUDIES- FARMERS ARE ALREADY DELIVERING FOOD AND FUEL WITH REAL SUSTAINABILITY
1:9 Fuel Land to Food Land Ratio
A 330 hectare farm in the Land Network farmers' group (Land Network Gainsborough, see above at page 143) has delivered taking a range of municipal and industrial "wastes" to make compost, so eliminating the use of mineral fertilisers, to grow good crops safely and these include oil seed rape which is used, on the same farm, to produce biodiesel to EN14214. They calculate that taking 100 hectares of oil seed rape grown this way will produce enough energy to run a farm of 1000 hectares, including all the field work and all the houses of the families who work that land. If this farm had those 1000 hectares, the remaining 900 hectares would, with the soils on that farm, produce 8,000 to 10,000 tonnes of wheat. Land Network Gainsborough can and will offer to supply biofuels at prices linked to gate fees on wastes used as their feedstocks.

One Million Loaves of Bread

Another farm in the Land Network group (Land Network Melton) does, again, use "wastes" to make compost to fertilise their land and eliminate groundwater pollution. The river Eye runs through their 330 hectare (800 acre) farm and the two farming brothers are involved with the river authority including conservation of water voles, freshwater crayfish and otters, plus the RSPB with avian biodiversity (76 bird species and 18 butterfly species) on the whole of their farm. They grow several crops. One is oil seed rape which they use to produce PPO, Pure plant Oil, and drive their own diesel engines with it. The wheat they produce would make one million loaves of bread.

Reversing Global Warming

The performances of the two farms above are directly related to environmental care, the reversal of global warming and long term sustainability.

Growing one hectare of oil seed rape and producing biofuels from wastes by the routes these farms use, will remove around 69 tonnes of Carbon dioxide from the atmosphere and pump around 73 tonnes of Oxygen back in. It is possible to argue about the figures in detail, but not about the principle.

What is described in this book is a way of making reversing global warming profitable and, therefore, sustainable. Proximity recycling of wastes, through the soil and green leaved crops, to produce biofuels works.

Appendix 1
References and Further Reading

1. Walker G & King D, The Hot Topic, Bloomsbury Publishing Plc, 2008.
2. Gore A, An Uncomfortable Truth, DVD, Paramount Pictures, 2006.
3. Menzies G, 1421 – The Year China Discovered the World, p300, Bantam Books, 2002.
4. Stanton W, The Rapid Growth of Human Populations 1750-2000, Multi-Science Publishing Company Ltd, 2003.
5. Daily Telegraph, 6th December 2007.
6. Dr Athol Klieve, Kangaroo Bacteria Could Fight Climate Change, Queensland Government.
7. Sara F Wright and Kristine A Nichols are with the SDAARS. This research is part of Soil Resource Management and ARS National Program (#202) described on the World Wide Web at www.nps.ars.usda.gov.
8. Ed Berg B et al, Plant litter, decomposition, humus formation, Carbon sequestration, Springer, 2003. This is one example of many hundreds of references to organic matter degradation and formation of humus in the academic literature and on the web. Empirical experience in the Land Network consortium of farmers who recycle wastes to land. Some of this experience is published, much is held in Codes of Practice, exclusive to the Network.
9. DTI unpublished reports by Land Network International Ltd under the Enterprise Initiative Programme.
10. ICI Plant Protection, as was, pursued their market for Gramoxone with both commercial (including Land Network International Ltd) and academic enthusiasm

in the 1980's with many publications and support trials, including by universities.

11. Butterworth B, How the Closed Loop Delivers True Sustainability, EnAgri, Issue 30 p27, September 2008.
12. Millward R & Robinson A, Upland Britain, David and Charles London (Publishers) Ltd, 1980.
13. Research by Land Network International Ltd, published later in Biofpr, December 2008.
14. Butterworth B, Clamping Down on Compost, Resource: American Society of Agricultural and Biological Engineers, April 2006.
15. Butterworth B, Nitrate Nonsense, Landwards; Institution of Agricultural Engineers, Early Summer, 2002.
16. Butterworth B, Waste in the Next Millennium, Resource – American Society of Agricultural and Biological Engineers, July 1997, p11-12.
17. Butterworth B, Reversing Global Warming, ReFocus, September/October 2006.
18. Ed Berg B et al, Plant litter, decomposition, humus formation, Carbon sequestration, Springer, 2003. This is one example of many hundreds of references to organic matter degradation and formation of humus in the academic literature and on the web.
19. Sara F Wright and Kristine A Nichols are with the SDAARS. This research is part of Soil Resource Management and ARS National Program (#202) described on the World Wide Web at www.nps.ars.usda.gov.
20. Kinsey N, Hands on Agronomy, Acres USA, 1999.
21. Chaudhry TM, Biogeochemical characterization of Metalliferous wastes and potential of arbuscular mycorrhizae in their phytoremediation. School of Science, Food and Horticulture, College of Science, Technology and Environment, University of Western

Sydney, Locked Bag 1997, Penrith South, DC NSW 1997, Austrailia.

22. Jeffries P, The Contribution of Mycorrhizal Fungi in Sustainable Maintenance of Plant Health and Fertility, Biology and Fertility of Soils, 2003, Vol37 p1-16.

23. Harrison MJ, Signalling in the Arbuscular Mycorrhizal Symbiosis, Annual Review of Microbiology, 2005, 59 p19.

24. Rose SC et al, The Design of a Pesticide and Washdown Facility, British Crop Protection Council Symposium, November 2001.

25. Butterworth B, The Straw Manual, Spon 1986.

26. DTI unpublished, but available, reports by Land Network International Ltd under the Enterprise initiative Programme.

27. Gordon Spoor, was a lecturer and researcher at the National College of Agriculture Engineering, Silsoe, UK, for more than 20 years in the 1980's and 1990's. He was acknowledged as a world expert on soil strengths and published many papers on the subject.

28. Albrecht W A, The Albrecht Papers, Charles Walter Books, 1919-1970.

29. Butterworth B, Managing the Soil Rumen and Ion Exchange, Arable Farming, 9 September 2000.

30. Defra, The Safe Sludge Matrix, www.defra.gov.uk/farm/waste/sludge/index.htm.

31. van Hoof TB et al, Forest Re-Growth on Medieval Farmland After the Black Death Pandemic - Implications for Atmospheric CO_2 Levels, Laboratory of Palaeobotany and Palynology, Department of Biology Faculty of Science, Utrecht, Elsevier, 2 February 2006, available on line.

32. Materials and Energy from Municipal Waste, Office of Technology Assessment, United States, Diane Publishing, 1979.

33. Turner C P and Carlile W R, Microbial Activity in Blocking Composts. 3. Degradation of Formaldehyde, ISHS Acta Horticulturae 150: International Symposium on Substrates in Horticulture other than Soils in Situ.
34. Butterworth B, Biofuels from Waste, ReFocus, May/June 2006.
35. Butterworth B, Reversing Global Warming, ReFocus September/October 2006.
36. Kumar Singh A et al, Assessment of the Input of Landfill in Groundwater Supply, Environment Monitor, 2008, 141; 309-321.
37. Environment Agency, Understanding Rural Land Use, NE-1/100-3.SK-C-BEKC (undated).
38. Defra, Waste Strategy Fact Sheets. www.defra.gov.uk/environment/waste/strategy/fact sheets/energy.htm
39. Evans T, personal communications with Land Network.
40. Tompkins P & Boyd C (1989), Secrets of the Soil, Harper and Row, New York.
41. There are many references in many places including on the web, for example at Washington State University, Cornell University and others.
42. Ryckeboer J, Microbiological Aspects of Biowaste During Composting in a Monitored Compost Bin, Journal of Applied Microbiology 94 (1), 2003.
43. Black is the New Green, Sequestration, Vol 442, p624-626, 10 August 2006, Nature Publishing Group.
44. Mann CC, Our Good Earth, National Geographic, September 2008, p80-107.
45. Turner C P and Carlile W R, Microbial Activity in Blocking Composts. 3. Degradation of Formaldehyde, ISHS Acta Horticulturae 150: International Symposium on Substrates in Horticulture other than Soils in Situ.
46. Evans T, personal communications with Land Network.

47. Rejon A et al, Mycorrhizal Fungi Influence Competition in a Wheat-Ryegrass Association Treated with Herbicide Diclifop, Applied Ecology, 1997, 7, p51-57.
48. Butterworth B, Far Eastern Agriculture, September/October 1987 and September/October 2003.
49. Steger K, Development of Compost Maturity and Actinobacteria Populations During Full-Scale composting of Organic Household Waste, Journal of Applied Microbiology, 2007.
50. Materials and Energy from Municipal Waste, Office of Technology Assessment, United States, Diane Publishing, 1979.
51. Collins J, Good to Great, Random House Business Books, 2001.
52. Kidney D, MP, Global Population Growth, debate in British Parliament on 4th February 2009.
53. Kabata-Pendia A & Mukherjee AB, Trace Elements from Soil to Human, Environmental Toxicology, 2007.
54. Freeman HM & Harris EF, Harzardous Waste Remediation: Innovative Treatment Technologies, CRC Press, 1995.
55. The Fourth Assessment Report, Intergovernmental Panel on Climate Change, United Nations 2007.

Appendix 2
Values of Materials

Chemical Value
A basic chemical analysis which will give a guide to value is set out below. However, first of all, remember the most important; Carbon. It is rare that this is itemised on a laboratory analysis. Carbon in organic form is the "sugar you put on your cereals at breakfast". It is the energy source for the micro-organisms in the compost and in the soil. Long chain Carbon molecules are the basis of the compounds which hold onto the nutrients such as Nitrogen, Phosphate and potash, and all the things in the list below which the crop needs, and stops them being leached to groundwater but allows the crop to be fed on them via the mycorrhiza.

- Nitrogen
- Phosphorous
- Potassium
- Calcium
- Sulphur
- Zinc
- Copper
- Nickel,
- Magnesium
- Sodium
- Cobalt

And a further list of micro-traces including
- Molybdenum
- Lithium
- And more, even including elements such as Arsenic

Physical Value

Organic matter will give a soil better properties in gas exchange and water movement and retention. Humus (that "Dirty Black Stuff") is part of this, but so is organic matter in general. Indeed, un-decomposed woody bits up to 50mm across, and more, are a positive benefit to the physical properties of most field soils, especially on clays.

Biological Value

A balanced diet is needed by the soil micro-organisms in just about the same way as we do. The Carbon is very important, too. If there is enough in the soil, it will raise the biological activity in the soil. This, in turn, will feed the crop more efficiently and faster, thus giving better crops (as happens in the Fen soils, which we can now copy).

Appendix 3
Materials successfully recycled to land by Land Network farms between 1995 and 2009
(It could and should have been a lot more.)

01	WASTES RESULTING FROM EXPLORATION, MINING, QUARRYING AND PHYSICAL AND CHEMICAL TREATMENT OF MINERALS
01 01	Wastes from mineral excavation
01 01 02	Wastes from mineral non-metalliferous excavation
01 04	Wastes from physical and chemical processing of non-metalliferous minerals
01 04 08	Waste gravel and crushed rocks other than those mentioned in 01 04 07
01 04 09	Waste Sands and Clays
02 `	WASTES FROM AGRICULTURE, HORTICULTURE, AQUACULTURE, FORESTRY, HUNTING AND FISHING, FOOD PREPARATION AND PROCESSING
02 01	Wastes from agriculture, horticulture, aquaculture, forestry, hunting and fishing
02 01 01	Sludges from washing and cleaning
02 01 03	Plant-tissue waste
02 01 06	Animal faeces, urine and manure (including spoiled straw), effluent, collected separately and treated off-site
02 01 07	Wastes from forestry
02 01 99	Wastes not otherwise specified
02 02	Wastes from the preparation and processing of meat, fish and other foods of animal origin
02 02 01	Sludges from washing and cleaning
02 02 04	Sludges from on-site effluent treatment
02 03	Wastes from fruit, vegetables, cereals, edible oils, cocoa, coffee, tea and tobacco preparation and processing; conserve production; yeast and yeast extract production, molasses preparation and fermentation
02 03 01	Sludges from washing, cleaning, peeling, centrifuging and separation
02 03 04	Materials unsuitable for consumption or processing
02 03 05	Sludges from on-site effluent treatment
02 04	Wastes from sugar processing
02 04 01	Soil from cleaning and washing
02 04 02	Off-specification calcium carbonate
02 04 03	Sludges from on-site effluent treatment
02 05	Wastes from the dairy products industry
02 05 01	Materials unsuitable for consumption or processing
02 05 02	Sludges from on-site effluent treatment
02 06	Wastes from the baking and confectionery industry
02 06 01	Materials unsuitable for consumption or processing

02 06 03	Sludges from on-site effluent treatment
02 07	**Wastes from the production of alcoholic and non-alcoholic beverages (except coffee, tea and cocoa)**
02 07 01	Wastes from washing, cleaning and mechanical reduction of raw materials
02 07 02	Wastes from spirits distillation
02 07 04	Materials unsuitable for consumption or processing
02 07 05	Sludges from on-site effluent treatment
03	**WASTES FROM WOOD PROCESSING AND THE PRODUCTION OF PANELS AND FURNITURE, PULP, PAPER AND CARDBOARD**
03 01	**Wastes from wood processing and the production of panels and furniture**
03 01 01	Waste bark and cork
03 01 05	Sawdust, shavings, cuttings, wood, particle board and veneer other than those mentioned in 03 01 04
03 03	**Wastes from pulp, paper and cardboard production and processing**
03 03 01	Waste bark and wood
03 03 02	Green liquor sludge (from recovery of cooking liquor)
03 03 05	De-inking sludges from paper recycling
03 03 07	Mechanically separated rejects from pulping of waste paper and cardboard
03 03 08	Wastes from sorting of paper and cardboard destined for recycling
03 03 09	Lime mud waste
03 03 10	Fibre rejects, fibre-, filler- and coating-sludges from mechanical separation
03 03 11	Sludges from on-site effluent treatment other than those mentioned in 03 03 10
04	**WASTES FROM THE LEATHER, FUR AND TEXTILE INDUSTRIES**
04 01	**Wastes from the leather and fur industry**
04 01 01	**Fleshings and lime split wastes**
04 01 05	**Tanning liquor free of chromium**
04 01 07	**Sludges, in particular from on-site effluent treatment free of chromium**
04 01 08	**Waste tanned leather (blue sheetings, shavings, cuttings, buffing dust) containing chromium**
04 01 09	**Wastes from dressing and finishing**
04 02	**Wastes from the textile industry**
04 02 20	Sludges from on-site effluent treatment other than those mentioned in 04 02 19
04 02 21	Wastes from unprocessed textile fibres
04 02 22	Wastes from processed textile fibres
06	**WASTES FROM INORGANIC CHEMICAL PROCESSES**
06 03	**Wastes from the MFSU of salts and their solutions and metallic oxides**
06 03 14	Solid salts and solutions other than those mentioned in 06 03 11 and 06 03 13

06 05	Sludges from on-site effluent treatment
06 05 03	Sludges from on-site effluent treatment other than those mentioned in 06 05 02
06 13	Wastes from inorganic chemical processes not otherwise specified
06 13 03	Carbon black
06 13 99	Wastes not otherwise specified
07	**WASTES FROM ORGANIC CHEMICAL PROCESSES**
07 02	Wastes from the MFSU of plastics, synthetic rubber and manmade fibres
07 02 12	Sludges from on-site effluent treatment other than those mentioned in 07 07 11
07 07 99	Wastes not otherwise specified
08	**WASTES FROM THE MANUFACTURE, FORMULATION, SUPPLY AND USE (MFSU) OF COATINGS (PAINTS, VARNISHES AND VITREOUS ENAMELS), ADHESIVES, SEALANTS AND PRINTING INKS**
08 01	Wastes from MFSU and removal of paint and varnish
08 01 99	Wastes not otherwise specified
08 03	Wastes from MFSU of printing inks
08 03 15	Ink sludges other than those mentioned in 08 03 14
08 03 99	Wastes not otherwise specified
08 04	Wastes from MFSU of adhesives and sealants (including waterproofing products
08 04 12	Adhesive and sealant sludges other than those mentioned in 08 04 11
08 04 14	Aqueous sludges containing adhesives or sealants other than those mentioned in 08 04 13
08 04 99	Wastes not otherwise specified
10	**WASTES FROM THERMAL PROCESSES**
10 01	Wastes from power stations and other combustion plants (except 19)
10 01 01	Bottom ash, slag and boiler dust (excluding boiler dust mentioned in 10 01 04)
10 01 03	Fly ash from peat and untreated wood
10 13	Wastes from manufacture of cement, lime and plaster and articles and products made from them
10 13 04	Wastes from calcination and hydration of lime
15	**WASTE PACKAGING; ABSORBENTS, WIPING CLOTHS, FILTER MATERIALS AND PROTECTIVE CLOTHING NOT OTHERWISE SPECIFIED**
15 01	Packaging (including separately collected municipal packaging waste)
15 01 01	Paper and cardboard packaging
15 01 03	Wooden packaging
15 01 09	Textile packaging
15 02	Absorbents, filter materials, wiping cloths and protective clothing

15 02 03	Absorbents, filter materials, wiping cloths and protective clothing other than those mentioned in 15 02 02
16	WASTES NOT OTHERWISE SPECIFIED IN THE LIST
16 10	Aqueous liquid wastes destined for off-site treatment
16 10 02	Aqueous liquid wastes other than those mentioned in 16 10 01
17	CONSTRUCTION AND DEMOLITION WASTES (INCLUDING EXCAVATED SOIL FROM CONTAMINATED SITES)
17 01	Concrete, bricks, tiles and ceramics
17 01 01	Concrete
07 01 02	Bricks
07 01 03	Tiles and ceramics
17 02	Wood, glass and plastic
17 02 01	Wood
17 04	Metals (including their alloys)
17 04 07	Mixed metals
17 05	Soil (including excavated soil from contaminated sites), stones and dredging spoil
17 05 04	Soil and stones other than those mentioned in 17 05 03
17 05 06	Dredging spoil other than those mentioned in 17 05 05
17 08	Gypsum-based construction material
17 08 02	Gypsum-based construction materials other than those mentioned in 17 08 01
18	WASTES FROM HUMAN OR ANIMAL HEALTH CARE AND/OR RELATED RESEARCH
18 01	Wastes from natal care, diagnosis, treatment or prevention of diseases in humans
18 01 04	Wastes whose collection and disposal is not subject to special requirements in order to prevent infection (for example dressings, plaster casts, linen, disposable clothing, diapers)
18 02	Wastes from research, diagnosis, treatment or prevention of disease involving animals
18 02 03	Wastes whose collection and disposal is not subject to special requirements in order to prevent infection
19	WASTES FROM WASTE MANAGEMENT FACILITIES, OFF-SITE WASTE WATER TREATMENT PLANTS AND THE PREPARATION OF WATER INTENDED FOR HUMAN CONSUMPTION AND WATER INTENDED FOR INDUSTRIAL USE
19 05	Wastes from aerobic treatment of solid wastes
19 05 01	Non-composted fraction of municipal and similar wastes
19 05 02	Non-composted fraction of animal and vegetable wastes
19 05 03	Off-specification compost
19 05 99	Wastes not otherwise specified
19 06	Wastes from anaerobic treatment of waste
19 06 03	Liquor from anaerobic treatment of municipal waste
19 06 04	Digestate from anaerobic treatment of municipal waste
19 06 05	Liquor from anaerobic treatment of animal and vegetable waste
19 06 06	Digestate from anaerobic treatment of animal and vegetable

	waste
19 07	**Landfill Leachate**
19 07 03	Landfill leachate other than those mentioned in 19 07 02
19 08	**Wastes from waste water treatment plants not otherwise specified**
19 08 02	Waste from desanding
19 08 05	Sludges from treatment of urban waste water
19 08 12	Sludges from biological treatment of industrial waste water other than those mentioned in 19 08 11
19 08 14	Sludges from other treatment of industrial waste water other than those mentioned in 19 08 13
19 09	**Wastes from the preparation of water intended for human consumption or water for industrial use**
19 09 01	Solid waste from primary filtration and screenings
19 09 02	Sludges from water clarification
19 12	**Wastes from the mechanical treatment of waste (for example sorting, crushing, compacting, pelletising) not otherwise specified**
19 12 01	Paper and cardboard
19 12 07	Wood other than those mentioned in 19 12 06
19 12 08	Textiles
19 12 09	Minerals (for example sand, stones)
20	**MUNICIPAL WASTES (HOUSEHOLD WASTE AND SIMILAR COMMERCIAL, INDUSTRIAL AND INSTITUTIONAL WASTES) INCLUDING SEPARATELY COLLECTED FRACTIONS)**
20 01	**Separately collected fractions (except 15 01)**
20 01 01	Paper and cardboard
20 01 08	Biodegradable kitchen and canteen waste
20 01 10	Clothes
20 01 11	Textiles
20 01 38	Wood other than that mentioned in 20 01 37
20 02	**Garden and park wastes (including cemetery waste)**
20 02 01	Biodegradable waste
20 02 02	Soil and stones
20 03	**Other municipal wastes**
20 03 02	Waste from markets

INDEX

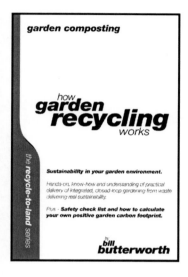

Lightning Source UK Ltd.
Milton Keynes UK
02 November 2009

145724UK00001B/47/P